AMERICAN JOURNAL OF MATHEMATICAL AND MANAGEMENT SCIENCES

VOL. 6, NOS. 3 & 4 1986

CONTENTS

This volume is

American Journal of Mathematical and Management Sciences, Volume 6 (1986), Nos. 3 & 4 (see front cover for cataloging information)

and is also issued as

Vehicle Routing with Time-Window Constraints: Algorithmic Solutions, Edited by Bruce L. Golden and Arjang A. Assad (see back cover for cataloging information).

AMERICAN JOURNAL OF MATHEMATICAL AND MANAGEMENT SCIENCES

VEHICLE ROUTING WITH TIME-WINDOW CONSTRAINTS

Bruce L. Golden and Arjang A. Assad
University of Maryland
College Park, Maryland 20742

The movement of goods from plants to warehouses and the delivery of products from such warehouses to the customers are typical examples of the distribution function within a firm producing goods. The last step in this flow of materials generally involves local transportation or delivery and is characterized by the distribution of products from a few outlets to a large number of customers. This activity alone could account for a large fraction of the total distribution-related costs of the firm. In recent years, an increasing amount of effort has focused on carrying this distribution out effectively so as to minimize the capital and operating costs of the firm.

In operational terms, a distribution manager must translate the customer orders that are ready for shipment into an effective set of schedules for vehicles and crews at his dis-

Key Words and Phrases: Vehicle Routing; Combinatorial Optimization.

1986, VOL. 6, NOS. 3 & 4, 251-260
0196-6324/86/030251-10 $5.00

posal. Each schedule specifies a set of customers to be visited by a vehicle and the associated deliveries (or pickups). It is not hard to find examples of the daily distribution problem involving 10-20 vehicles and 100-500 customers. In general, the construction of schedules is a complicated task due to the size and mix of the fleet, the number of customers involved, and the fact that such schedules must be put together in a relatively short period of time. For problems of even moderate size, say, with over 5 vehicles and 50 deliveries, it may be very difficult for a human to construct efficient schedules that satisfy all of the many real-world constraints placed on the deliveries. For this reason, the use of computerized routing and scheduling techniques has assumed increasing practical importance and relevance in the last five years.

Past studies and successful implementation efforts have demonstrated that the use of routing and scheduling models and their associated solution techniques is instrumental in realizing important distribution-related cost savings. For example, an on-line computerized routing and scheduling system of a large supplier of industrial gases (see Bell, Dalberto, Fisher, Greenfield, Jaikumar, Kedia, Mack, and Prutzman (1983)) has been saving 6% to 10% of operating costs annually. Computerized vehicle routing reduced yearly delivery costs by over 15% for a medical equipment supplier (see Fisher, Greenfield, Jaikumar, and Lester (1982)). Annual savings of over 13% are reported by Harrison and Wills (1983) as a result of computerized milk collection in Ireland. In addition, algorithmic routing techniques have led to reductions of 13% per year in total distance covered in a "meals on wheels" application in Atlanta, Georgia (see Bartholdi, Platzman, Collins, and Warden (1983)).

The increasing awareness of the savings potential of routing and scheduling algorithms among firms engaged in distribution activities and successful implementations of routing systems have led to greater interest in devising solution techniques for handling realistic but complicated constraints.

This Special Issue focuses on one such class of constraints, known as time-window constraints, that have to do with the timing of visits to customer sites. In what follows, we briefly describe the basic vehicle routing problem and the role of time-window constraints in this problem.

Consider a set of customers with known demands who are supplied from a single depot using an operating fleet of vehicles. All vehicles have the same capacity and are available for a fixed number of hours each day (eight hours, say). The basic vehicle routing problem involves constructing a set of delivery routes, based at the depot, that supply all customer requirements using minimum travel distance. This problem is illustrated in Figure 1. In this figure, there are three vehicles each with a capacity of thirty units; there are a total of nine customers, each with a demand of ten units.

In realistic distribution environments, the preceding basic problem becomes more complex and involved very quickly. Complications may arise due to a heterogeneous fleet of vehicles, pick-up or delivery time windows at the customer locations, backhaul considerations, overtime decisions, multiple daily routes for a vehicle, open and close times at the customer locations, and the presence of more than one central depot, or multiple commodities.

The June 1983 issue of COMPUTERS & OPERATIONS RESEARCH, which is devoted entirely to vehicle routing and scheduling, describes how these complications arise in greater detail. In this Special Issue of the AMERICAN JOURNAL OF MATHEMATICAL AND MANAGEMENT SCIENCES, we focus on time windows and algorithmic techniques for solving vehicle routing problems in the presence of such time-window constraints.

A time window at a customer location is a pre-specified time interval within which the customer must, or desires to, be visited. For example, the customer may require that his delivery take place between 9 and 11 a.m. In some cases, multiple windows may be specified as in "deliver between 8 and 11 a.m. or

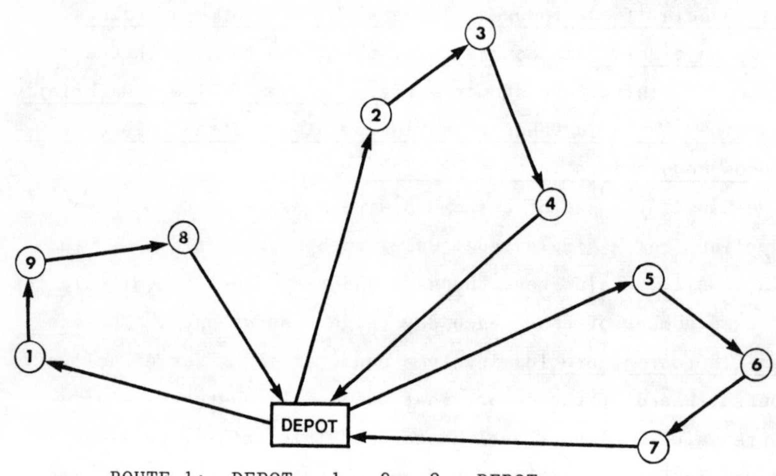

ROUTE 1: DEPOT - 1 - 9 - 8 - DEPOT

ROUTE 2: DEPOT - 2 - 3 - 4 - DEPOT

ROUTE 3: DEPOT - 5 - 6 - 7 - DEPOT

FIGURE 1. Basic Vehicle Routing Problem.

2 and 4 p.m." To see how time windows can complicate a routing problem, it is useful to examine the simple example shown in Figure 2. Assume that each of the three customers requires 1/3 of a truckload and that travel times between locations are as marked on the edges (in hours). The time window for each of customers 1 and 3 is 9 a.m. to noon, while customer 2 must be serviced after noon. In the absence of time windows, the minimum travel time route is D - 1 - 2 - 3 - D, where D denotes the depot. However, this route is not feasible when time windows are imposed and the new optimal set of routes is D - 1 - 2 - D and D - 3 - D. In this example, the imposition of time windows increased the total travel time of the optimal solution as well as the number of vehicles required.

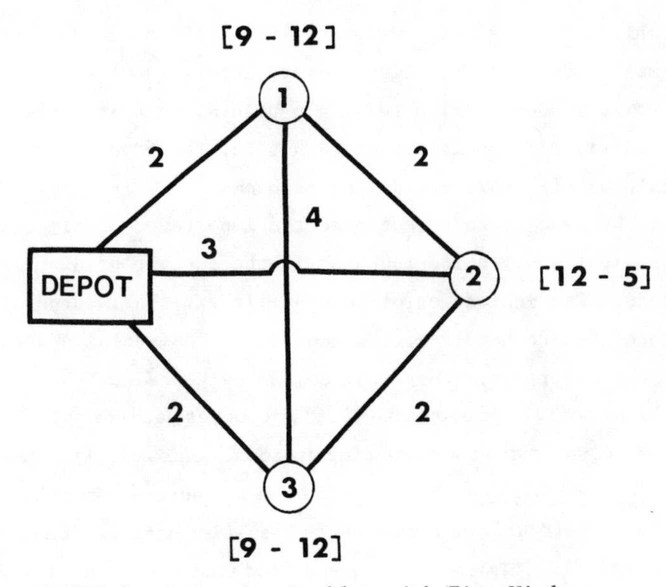

FIGURE 2. A Routing Problem with Time Windows.

Time-window constraints occur widely and frequently in the practical routing of vehicles. In modeling time windows, one can distinguish between a hard window (which restricts the delivery time at the customer site), and a soft window (where the window can be missed, whereupon a penalty based on the deviation is incurred). For example, open and close times are generally viewed as hard windows since the customer site is unable to take delivery prior to its open time or after its close time. More generally, unless the customer facility is completely unavailable or inaccessible outside the specified time window, the window may be viewed as being soft. Of course, even hard time windows can be treated algorithmically as soft windows with very large penalty terms. In that case, the penalty-free solution to the problem, if found, would meet all window requirements. The other factor guiding the modeling and algorithmic approach is the "tightness" of the time windows associated with a particular problem. In some product distribution activities, the time windows span large portions of the working day. In other problems, such as package or mail collection and delivery, the time windows span relatively short intervals of time. Indeed, as time windows shrink to single points, so that a precise time of service is specified for each site, the problem assumes the features of a pure scheduling problem.

We have already mentioned the importance of time windows in practical applications such as in the context of distributing goods. One indication of this utility is the incorporation of procedures to handle time windows into commercial systems aimed at the distribution marketplace, as evidenced by the survey of Golden, Bodin, and Goodwin (1986). However, other applications areas also involve and use time windows. One well-known example is the "Dial-a-Ride" problem, where a set of customers must be transported from their origin locations to specified destination points. In this problem, time windows are associated with the pickup (at the origin) and the delivery (to the destination) of each customer. A "shared-cab" system exhibits these features in the private sector, while transportation services provided to the elderly and the handicapped serve as a good public sector example. Sexton and Bodin (1985) describe the latter application and the results of their implementation of a routing and scheduling system. While the dial-a-ride problem and the pickup and delivery problem with time windows have a common structure, the former generally involves much tighter time windows (people dislike waiting or being delayed).

The papers collected in this issue discuss time windows in the context of basic vehicle routing, pickup and delivery, and dial-a-ride problems. Time windows occur in a number of other routing and scheduling problems. In school bus scheduling, for example, time windows are associated with students' arrival at, or departure from, each school. An overview of this complex problem can be found in Bodin, Golden, Assad, and Ball (1983). Another topic related to time windows but not included in this issue is assignment routing. In this problem, the frequencies of visits to customers are specified and the routing algorithm must assign these visits to various days of the week (or a longer planning cycle). This problem may be viewed as lengthening the time windows to span a full day while the planning horizon is simultaneously stretched over several days. Beltrami and Bodin (1974) describe an instance of this problem in waste collection. Algorithms designed specifically for assignment routing appear in Russell and Igo (1979) and Christofides and Beasley (1984). A related problem is addressed by Golden, Assad, and Dahl (1984).

The preceding comments should indicate the importance and pervasiveness of time windows in routing problems. In this issue, five articles by leading researchers are brought together, all dealing with time windows in the context of routing and scheduling.

The first paper, by Baker and Schaffer, examines the basic vehicle routing problem with hard time windows. Given the vast literature on the vehicle routing problem in the absence of time window constraints, this is a natural and practical extension. The paper proposes and tests insertion and improvement heuristics to handle the added window constraints.

The second and third papers both study the dial-a-ride problem, but treat different variants of this problem. Desrosiers, Dumas, and Soumis address a single-vehicle problem with hard time windows on both pickups and deliveries. They show that the size of the state space of a suitably defined dynamic programming formulation of this problem can be reduced significantly by means of appropriate rules. This allows the authors to solve problems with 40 customers to optimality with dynamic programming. Psaraftis, in the third paper, considers a multi-vehicle case in which a time window is specified for each pick-up or delivery, but not both.

Two heuristic procedures are described and compared. One procedure assumes the windows are hard, and the other treats them as being soft. The paper discusses computational results with some large problems (one has over 2000 customers). It also provides many insights into this class of dial-a-ride problems.

In the fourth paper, Sexton and Choi extend some of the methodology developed earlier for dial-a-ride problems to handle general pickup and delivery problems with (soft) time windows. As in the second paper, a single vehicle problem with time windows on both pickups and deliveries is treated. The objective is to minimize the total duration of the vehicle schedule. The solution technique involves specially designed heuristics embedded within Benders' decomposition. The fact that this general approach can be used effectively for both dial-a-ride problems, as in Sexton and Bodin (1985), and pickup and delivery problems as in the current paper, shows the close structural similarity of these two problem classes.

In the fifth and final paper, Solomon addresses the minimal spanning tree problem with time windows. This problem may appear to be an easier time-window problem, but is actually quite difficult, as Solomon shows in this paper. Nevertheless, Solomon argues that it can be useful as a subproblem in more general routing problems and he, therefore, develops and tests greedy and insertion heuristics to solve it. Solomon's paper and the earlier work of Desrosiers, Pelletier, and Soumis (1983) are two examples of work on time windows imposed on combinatorial structures other than a set of tours--spanning trees in the former paper, and shortest paths in the latter.

As guest editors, we take pleasure and some pride in the fact that this Special Issue is the first collection of articles focusing specifically on the theme of time windows in routing. The papers chosen for this issue reflect a variety of modeling and algorithmic approaches to routing problems with time windows. Although most of the research is of very recent vintage, this body of work already marks the significant progress made in handling realistic time-window problems. We feel confident that the techniques presented in this issue will not only continue to develop, but will also motivate and contribute to work on other complicated routing problems arising in practice.

REFERENCES

Bartholdi, J., Platzman, L., Collins, R., and Warden, W. (1983). A minimal technology routing system for meals on wheels. Interfaces, 13(3), 1-8.

Bell, W., Dalberto, L., Fisher, M., Greenfield, A., Jaikumar, R., Kedia, P., Mack, R., and Prutzman, P. (1983). Improving the distribution of industrial gases with an on-line computerized routing and scheduling optimizer. Interfaces, 13(6), 4-23.

Beltrami, E. and Bodin, L.D. (1974). Networks and vehicle routing for municipal waste collection. Networks, 4, 65-94.

Bodin, L. D., Golden, B., Assad, A., and Ball, M. (1983). The state of the art in the routing and scheduling of vehicles and crews. Computers & Operations Research, 10, 63-211.

Christofides, N. and Beasley, J. (1984). The period routing problem. Networks, 14, 237-256.

Desrosiers, J., Pelletier, P., and Soumis, F. (1983). Plus court chemin avec contraintes d'horaires. RAIRO, 17, 357-377.

Fisher, M., Greenfield, A., Jaikumar, R., and Lester, J. (1982). A computerized vehicle routing application. Interfaces, 12(4), 42-52.

Golden, B., Assad, A., and Dahl, R. (1984). Analysis of a large scale vehicle routing problem with an inventory component. Large Scale Systems, 7, 181-190.

Golden, B., Bodin, L., and Goodwin, T. (1986). Microcomputer-based vehicle routing and scheduling software. Computers & Operations Research, 13, forthcoming.

Harrison, H. and Wills, D. (1983). Product assembly and distribution optimization in an agribusiness cooperative. Interfaces, 13(2), 1-9.

Russell, R. and Igo, W. (1979). An assignment routing problem. Networks, 9, 1-17.

Sexton, T. R. and Bodin, L. D. (1985). Optimizing single vehicle many-to-many operations with desired delivery times: I. Scheduling and II. Routing (two parts). Transportation Science, 19, 378-435.

Received 5/30/86.

AMERICAN JOURNAL OF MATHEMATICAL AND MANAGEMENT SCIENCES

SOLUTION IMPROVEMENT HEURISTICS FOR THE VEHICLE ROUTING AND SCHEDULING PROBLEM WITH TIME WINDOW CONSTRAINTS

Edward K. Baker and Joanne R. Schaffer

Department of Management Science and
Computer Information Systems
University of Miami
Coral Gables, Florida 33124

SYNOPTIC ABSTRACT

Branch exchange techniques, such as the well-known 2-opt and 3-opt procedures, are among the most powerful heuristics available for the solution of the classic vehicle routing problem. The imposition of time window constraints on customer delivery time within the vehicle routing problem, however, introduces several complexities that can reduce the power of these techniques. In this paper, two test data sets for the vehicle routing and scheduling problem with time window constraints are studied. Initial solutions are obtained using a variety of heuristics. These solutions are then improved using branch exchange procedures modified to incorporate the time window constraints.

Key Words and Phrases: vehicle routing and scheduling; time window constraints; branch exchange heuristics.

1986, VOL. 6, NOS. 3 & 4, 261-300
0196-6324/86/030261-40 $11.00

1. INTRODUCTION.

The vehicle routing problem (VRP) involves the routing of a fleet of vehicles of known capacity from a central depot to a set of customers with known demands. All routes must originate and terminate at a central depot. Additionally, the total demand on any route must not exceed the capacity of the vehicle assigned to service that route. The objective of the VRP is typically to minimize the total cost or distance traveled over all routes.

The vehicle routing and scheduling problem with time window constraints (VRSPTW) is a variation of the vehicle routing problem in which each of the customers to be serviced may have one or more time windows during which the customer service must be scheduled. Vehicle routing problems with time window constraints are typically encountered in situations where the customer must provide access, verification, or payment upon delivery of the product or service. Additionally, vehicle routing problems with a time dependent inventory component may be formulated as vehicle routing problems with time window constraints. Examples of such problems include the delivery of home heating oil, the supply of liquified gases to hospitals, and the delivery of many other valuable products stored in bulk by the consumer. In the paratransit area, dial-a-ride services for the elderly, handicapped, or impaired, generally require the use of time windows on client pickup and delivery time.

Although the VRP has been studied extensively, see Bodin, Golden, Assad, and Ball (1983) for a recent comprehensive survey, the VRSPTW has only recently attracted much research attention. In work on the dial-a-ride problem, Sexton (1979) includes time window constraints as part of an integer programming model of the problem. In related work, Psaraftis (1980) proposes a dynamic programming approach to the dial-a-ride problem which incorporates time window constraints. Working with time window constrained traveling salesman problems, Christofides, Mingozzi, and Toth (1981) propose the use of a state space relaxation procedure for optimal solutions to this special case of the VRSPTW. Baker (1983) proposes the use of a disjunctive graph model for obtaining optimal solutions to tightly constrained time window problems. Desrosiers, Soumis, and Desrochers (1983) discuss the use of a column generation procedure for the solution of the VRSPTW with applications to school bus scheduling.

Although numerous heuristic procedures have been successfully applied to the VRP, few attempts at the solution of the VRSPTW using heuristics have been reported in the literature. Russell (1977) reports success with an M-tour procedure for vehicle routing problems with a few time window constraints. Solomon (1983a) provides a comprehensive survey of heuristic route construction procedures for the VRSPTW.

In this paper we extend Solomon's work on the VRSPTW to include the application of route improvement procedures to heuristically generated initial solutions. The solution improvement procedures considered in this computational study are variants of the 2-opt and 3-opt branch exchange procedures, well-known in the vehicle routing literature, adapted to include the vehicle capacity and the time window constraints of the VRSPTW. Computational results are presented for route construction and route improvement procedures for 25 test problems obtained from the VRSPTW literature. Additional computational experience is reported for a very large (250 customer) real world VRSPTW.

1.1. Mathematical Models for the VRSPTW. Traditionally the VRP has been modeled as a linear integer program. The various types of formulations have included vehicle flow based, set covering, and commodity flow based models. The recent paper of Magnanti (1981) provides an excellent survey of the various formulations.

Mathematical models proposed for the VRSPTW have generally been extensions of models for the VRP that explicitly include time window constraints. Examples of vehicle flow based formulations of the VRSPTW may be found in Golden, Magnanti, and Nguyen (1977), Sexton (1979), or Solomon (1983a). A time oriented formulation for the VRSPTW based upon a disjunctive graph model from scheduling theory is given by Baker (1983). Details of both types of formulations are presented in Appendix A below.

1.2. Computational Complexity of the VRSPTW. Since the algorithms involved in this study are heuristic in nature (i.e., optimality is not proved by the algorithms), the computational complexity of the VRSPTW is of some interest. The general VRSPTW has been shown to be NP-hard. A problem is NP-hard if it belongs to the class of problems for which no solution procedures of polynomial computational complexity are known to exist. (See Lenstra and Rinnooy Kan (1981)). It is clear that if this were not the case, the choice of a single vehicle of unlimited capacity and the relaxation of the time window constraints within the VRSPTW model would solve the traveling salesman problem.

Although the VRSPTW is NP-hard, the presence of tight time window or vehicle capacity constraints in a particular VRSPTW may restrict the computational effort required to solve that problem instance. Bell, Dalberto, Fisher, Greenfield, Jaikumar, Mack, and Prutzman (1983) have solved large-scale VRSPTW effectively by using a generalized assignment algorithm within a Lagrangian relaxation procedure. Results presented by Baker (1983) indicate that certain traveling salesman problems with tight time window constraints can be solved efficiently with a branch and bound procedure.

In the analysis of the performance of heuristic algorithms for the VRSPTW, Solomon (1983b) has shown the worst-case performance of several route construction heuristics for the VRSPTW to be arbitrarily poor. (This implies that the length of the vehicle tours produced by the heuristic can, in certain problem instances, be made arbitrarily longer than the optimal length.) The VRSPTW test problems used in this study, however, do have a number of special structures. It is expected, therefore, that the combination of the route construction and route improvement procedures applied to these problems will produce results that will be much better than their theoretical worst-case performance may suggest.

2. DESCRIPTION OF THE HEURISTIC ALGORITHMS.

2.1. Route Construction Algorithms. In performing the computational analysis of this study, four heuristic route construction algorithms were coded: a time oriented nearest neighbor procedure and three insertion procedures. Each of the procedures will be described briefly below. A more explicit discussion of these procedures may be found in Solomon (1983a).

The time oriented nearest neighbor procedure begins each new route at the depot and adds to the end of the route the unrouted customer that is nearest in terms of the particular measure defined for the algorithm. In the standard VRP, the measure used is generally the distance from the most recently routed customer to the unrouted customer. Routes are built sequentially, the first route being completed before the second route is begun.

To adapt this algorithm to the VRSPTW, two additional time measures are added. The first of these, $t(i,j)$, measures the time between the departure from customer i and the service of customer j. Given the end of the route is currently at customer i, the second measure, $p(i, j)$, is the time until customer j would be infeasible on the route. In a sense $p(i,j)$ is the urgency of customer j. These two measures may be defined mathematically as:

$$t(i,j) = \text{Max}\ (e(j),\ t(i) + s(i) + d(i,j)) - (t(i) + s(i)),$$

$$p(i,j) = f(j) - (t(i) + s(i) + d(i,\ j))$$

where: $d(i,j)$ = distance between customer i and customer j

$e(i)$ = the earliest time window of customer i

$f(i)$ = the latest time window of customer i

$s(i)$ = the service or dwell time at customer i

$t(i)$ = actual delivery time of customer i

Distance and time measures are then combined to measure the nearness of an unrouted customer j as

$$c(i,j) = A1 * d(i,j) + A2 * t(i,j) + A3 * p(i,j)$$

where: $A1 + A2 + A3 = 1$.
The values of the weighting coefficients used in this study were $A1 = A2 = A3 = .33$.

The three variations of the insertion heuristic now to be described all begin the construction of a route by selecting the unrouted customer furthest from the depot, i.e. node 0, to begin the route. Each unrouted customer is then considered in two steps. Initially, the optimal insertion position of the unrouted customer is determined, then the insertion decision criterion is calculated. The unrouted customer with the optimal value of the second insertion criterion is then selected to be inserted in the route.

The first insertion algorithm uses decision criteria based solely on distance measures. The first of these, C1, is equal to the increased distance on the route created by the insertion of unrouted customer u between routed customers i and j. The second criterion, C2, calculates the savings in making the insertion. Mathematically, these criteria are:

$$C1 = d(i,u) + d(u,j) - A1*d(i,j)$$

$$C2 = A2*d(0,u) - C1.$$

Our choice of $A1=1$ and $A2=2$ for this study allows the algorithm to perform in the same manner as the generalized savings criterion described by Gaskell (1967).

The second insertion algorithm introduces temporal aspects to the decision process. Allowing $t(j/u)$ to be the new arrival time at customer j given customer u has been inserted between i and j, the insertion position is determined by a weighted combination of increased distance and increased arrival time. Defining

$$C11 = d(i,u) + d(u,j) - A1*d(i,j)$$

and

$$C12 = t(j/u) - t(j)$$

then

$$C1 = A11*C11 + A12*C12,$$

where

$A11 + A12 = 1.$

The second insertion criterion, C2, is then defined as a weighted combination of new route distance, Rd, and route time, Rt. The second criterion may be expressed as:

$C2 = At*Rt + Ad*Rd$, where $At + Ad = 1$.

The values chosen here more or less arbitrarily for this insertion procedure were A1 = 1, A11 = A12 = .5, and At = Ad = .5.

The third insertion algorithm adds the additional consideration of waiting time to the temporal criteria of the second insertion procedure. The additional term, C13, equal to the waiting time at customer u, is added to the C1 criterion for the selection of the insertion position. This criterion is then

$C1 = A11*C11 + A12*C12 + A13*C13$

where:

$C13 = f(u) - t(u)$ and $A11 + A12 + A13 = 1$.

The value of the second insertion decision criterion, C2, for this third insertion procedure is set equal to C1. The values of weighing coefficients used in this study were A11=A12=A13=.33.

2.2. Route Improvement Algorithms. The route improvement heuristics used in this study are all variants of the k-opt branch exchange procedures proposed by Lin (1965) and Lin and Kernighan (1973). Branch exchange procedures attempt to discover an improved solution by exchanging k arcs or branches of the current routing with k branches not in the current routing.

In this study the branch exchange heuristics are applied in two distinct phases. In the first phase, the branch exchange heuristic is applied to possible exchanges within each route. In the second phase, exchanges between routes are examined. Each of these procedures is described briefly below.

2.2.1. First Phase: Within Route Improvement Procedures. The within route branch exchange heuristics include both a 2-opt and a 3-opt procedure. To improve computational efficiency the heuristics are implemented in a sequential manner, the 3-opt being applied to the 2-opt solution. Additionally, the procedures have been modified to check the feasibility of all vehicle capacity and time window constraints on the newly reconfigured route. A weighted combination of total route distance, Rd, and total route time, Rt, are used to compare the advantage of possible exchanges. In this study, the cost criterion used was:

Route Cost = Ad*Rd + At*Rt

where:

Ad = At = .5.

2.2.2. Second Phase: Between/Among Routes Improvement Procedures. The between route 2-opt branch exchange heuristic considers the exchange of two arcs on two different routes. This procedure has the effect of exchanging the beginning and ending portions of two distinct routes. This same procedure has been applied to vehicle scheduling problems by Russell (1977) and to airline crew scheduling problems by Baker, Bodin, Finnegan, and Ponder (1979). A weighted combination of total route time and distance, identical to that used in the within route exchange procedures, was used to test for reduced cost route reconfigurations here.

The between route 3-opt procedure allows a portion of route i to be inserted on route j. Whereas the routes in the classic VRP may be traversed in either direction without loss of generality, the routes constructed for the VRSPTW typically possess a definite orientation. These routes often become infeasible, i.e. violate the time window constraints, if traversed in their reverse order. In fact, the computational results of this study reveal that even the traversal of a short sequence of arcs in reverse order will often cause the route to become infeasible.

The final route improvement procedure investigated in this study was the 3-opt procedure among three routes. Given the time orientation of these original routes, this 3-opt among three routes procedure allows for each route to be bisected and then reconfigured with a portion of another of the selected routes. If the set of routes to be examined is 2-optimal, there are only two configurations of the "3-opt among three routes" procedure that must be examined. Although the number of possibilities that must be examined is restricted, a large problem with several routes will still present a formidable computational task.

3. COMPUTATIONAL RESULTS.

The computational results of this paper were obtained by applying the heuristic algorithms discussed in Section 2 to problems and data sets from the VRSPTW literature. Each of the test problems and data sets is discussed briefly below. The interested reader may find sample test problems and copies of the solution algorithms in Appendix B of this paper.

The first set of test problems used in this study is a collection of time window constrained traveling salesman problems. These ten problems, denoted B11 through B52, are described in detail in Baker (1983). Briefly, these problems are adapted from the vehicle routing problems found in Eilon, Watson-Gandy, and Christofides (1971). The time window constrained traveling salesman problems were formed by finding a nearest neighbor solution through the points of each problem and then placing time windows about the nearest neighbor arrival times so that no two time windows overlapped. A specified percentage of randomly selected time windows was then relaxed. Problem Bj1 represents problem j with 90% of the time windows in effect. Problem Bj2 represents problem j with 75% of the time windows in effect.

The heuristic solution procedures described in this paper were applied to the test problems in multiple phases. The solution process begins with the route construction heuristic. The route construction heuristic reads the initial data file of the problem and creates an initial solution file. It is noteworthy that since the number of vehicles used in the solution is not fixed a priori, a feasible solution will always be found if one exists. This initial set of routes is then exposed to the within route improvement procedures and a within route improved solution file is created. The within route improved solution file is then subjected to the between/among route improvement procedures and a between route improved solution file is formed. The improvement procedures are repeated until neither the within route nor the between route improvement procedures produce an improved solution.

Tables 1 and 2 present the results obtained by applying the heuristic algorithms described in this paper to the first set of test problems. The tables first identify each problem, its number of customers, and the percentage of customers that have time window constraints on their delivery time. Next, the optimal solution to these traveling salesman problems with time windows is given. These optimal solutions were obtained using the disjunctive graph model described in Baker (1983).

The computational results for each solution, i.e. the optimal and the initial and improved solutions for each of the heuristic procedures, are reported with four measures of solution quality. The four values in the column for each case correspond to the total time of the schedule (TT), the total distance traveled over all routes (TD), the number of vehicles required (NV), and the CPU time (CPU), in seconds, required to perform all of the heuristic procedures of that phase of the solution process on a UNIVAC 1100/82 mainframe. In the case of the heuristic improvement procedures, the total computation time is the sum of the times of all the required repetitions of the procedures. It is also noted that the difference between the total time over all routes and the total distance traveled is equal to the waiting time plus the customer service times. In this first set of test problems, the customer service times were assumed to be zero.

The second set of test problems used in this study was

Table 1. Computational Results for the Time Window Constrained Traveling Salesman Test Problems B11 - B31

Problem		B11	B12	B21	B22	B31
Customers		8	8	12	12	21
% Windows		90	75	90	75	90
Optimal Solution	TT:	6564	6384	162	162	354
	TD:	6564	6384	158	158	350
	NV:	1	1	1	1	1
	CPU:	.039	.036	.057	.089	.167
Nearest Neighbor Initial Solution	TT:	7770	6564	162	188	382
	TD:	7310	6564	158	184	377
	NV:	1	1	1	1	1
	CPU:	.0282	.0278	.0450	.0452	.1040
Nearest Neighbor Improved Solution	TT:	6564	6384	Initial	162	354
	TD:	6564	6384	Solution	158	350
	NV:	1	1	Optimal	1	1
	CPU:	.1600	.1066		.5288	7.440
Insert 1 Initial Solution	TT:	6564	6564	164	172	355
	TD:	6564	6564	164	172	355
	NV:	1	1	1	1	1
	CPU:	.0306	.0308	.0546	.0550	.1476
Insert 1 Improved Solution	TT:	Initial	6384	162	162	354
	TD:	Solution	6384	158	158	350
	NV:	Optimal	1	1	1	1
	CPU:		.1068	.5666	.7622	2.342
Insert 2 Initial Solution	TT:	6564	9116	164	164	355
	TD:	6564	7414	164	164	355
	NV:	1	2	1	1	1
	CPU:	.0432	.0356	.0664	.0676	.2016
Insert 2 Improved Solution	TT:	Initial	7532	162	162	354
	TD:	Solution	6255	158	158	350
	NV:	Optimal	2	1	1	1
	CPU:		.1740	.6012	.5994	3.644
Insert 3 Initial Solution	TT:	6564	6564	164	166	354
	TD:	6564	6564	164	162	350
	NV:	1	1	1	1	1
	CPU:	.0410	.0328	.0596	.0622	.1746
Insert 3 Improved Solution	TT:	Initial	6384	162	162	Initial
	TD:	Solution	6384	158	158	Solution
	NV:	Optimal	1	1	1	Optimal
	CPU:		.1068	.5946	.6252	

Legend: TT: Total Route Time
TD: Total Route Distance
NV: Number of Vehicles
CPU: CPU Seconds UNIVAC 1100/82

Best initial solution underlined
Best final solution boxed

Table 2. Computational Results for the Time Window Constrained Traveling Salesman Test Problems B32 - B52

Problem		B32	B41	B42	B51	B52
Customers		21	29	29	50	50
% Windows		75	90	75	90	75
Optimal Solution	TT:	354	4755	4755	5631	5632
	TD:	341	4753	4753	5624	5557
	NV:	1	1	1	1	1
	CPU:	2.729	.6810	15.98	1.926	40.03
Nearest Neighbor Initial Solution	TT:	431	5719	6514	6310	7331
	TD:	406	5501	5821	6296	7027
	NV:	1	1	1	1	1
	CPU:	.1048	.1784	.1786	.4630	.4630
Nearest Neighbor Improved Solution	TT:	354	4755	4755	5631	5632
	TD:	341	4753	4753	5624	5557
	NV:	1	1	1	1	1
	CPU:	10.47	39.71	48.12	271.8	1082.
Insert 1 Initial Solution	TT:	397	7304	7610	6325	8741
	TD:	389	6793	6865	6319	6274
	NV:	1	2	2	2	2
	CPU:	.1536	.2634	.2802	.9108	1.003
Insert 1 Improved Solution	TT:	354	4755	4759	5714	5632
	TD:	341	4753	4759	5714	5557
	NV:	1	1	1	2	1
	CPU:	6.183	13.73	17.27	132.3	420.0
Insert 2 Initial Solution	TT:	355	7896	9055	8634	6504
	TD:	355	5791	6674	7404	6416
	NV:	1	2	2	2	1
	CPU:	.2108	.4272	.4534	1.446	1.996
Insert 2 Improved Solution	TT:	354	4755	4755	5631	5632
	TD:	341	4753	4753	5624	5557
	NV:	1	1	1	1	1
	CPU:	4.053	50.24	72.17	99.54	833.9
Insert 3 Initial Solution	TT:	354	4755	5913	5636	7589
	TD:	342	4753	5636	5636	7285
	NV:	1	1	1	1	1
	CPU:	.1872	.3540	.3860	1.304	1.399
Insert 3 Improved Solution	TT:	354	Initial	4755	5631	5632
	TD:	341	Solution	4753	5624	5557
	NV:	1	Optimal	1	1	1
	CPU:	2.810		30.28	120.9	1091.

Legend:
TT: Total Route Time
TD: Total Route Distance
NV: Number of Vehicles
CPU: CPU Seconds UNIVAC 1100/82

Best initial solution underlined
Best final solution boxed

created by Professor Marius Solomon of Northeastern University (see Solomon (1983a) for details) for the multiple vehicle routing and scheduling problem with time windows. This second set of test problems consists of two distinct groups. The first group of 12 problems, denoted S1 - S12, was adapted from the 100 customer vehicle routing problems of Christofides, Mingozzi and Toth (1979) by randomly generating time windows for a fixed percentage (either 75 or 90 percent) of the customers. In this process, first a percentage of time windows was established, then the actual customers to receive time windows were chosen. Finally the time windows were determined. The determination of the time windows was made by randomly positioning the center of the window on the interval formed by the earliest possible and latest allowable arrival times of a vehicle traveling directly to and from the depot while remaining within the route start and end time constraints. The width of the window was then selected from a specified normal distribution.

The second group of this second set of test problems is a collection of multiple vehicle time window routing problems with a special "clustered" structure. In these problems, designated S13, S14, and S15 in this paper, each of ten customer clusters was considered to be a separate vehicle route. The route within each cluster was then optimized using a 3-opt procedure. The time windows within each cluster were then centered on the arrival time produced in the 3-opt solution. The width of each time window was then randomly selected from a user specified normal probability distribution.

The results for the second set of vehicle routing and scheduling problems with time window constraints are presented in Tables 3, 4, and 5. The format used in these tables is identical to that used for Tables 1 and 2 with one exception being that rather than an optimal solution being reported for these multiple vehicle problems, the best known solution is reported. This best known solution is the lowest of the equally weighted sum of total

Table 3. Computational Results for the Multiple Vehicle Routing and Scheduling Problem with Time Windows: Test Problems S1 - S5

Problem		S1	S2	S3	S4	S5
Customers		100	100	100	100	100
% Windows		75	90	75	90	75
Best Known Solution	TT:	3715	3562	2805	2476	2854
	TD:	1873	1705	1484	1188	1604
	NV:	21	19	14	11	15
	CPU:	2.7	2.8	3.0	3.4	2.7
Nearest Neighbor Initial Solution	TT:	4132	4063	3362	3145	3134
	TD:	2229	2058	1743	1464	1724
	NV:	23	20	15	14	14
	CPU:	.9282	.9278	.8450	.9580	.8670
Nearest Neighbor Improved Solution	TT:	3563	3296	2632	[2166]	2375
	TD:	1732	1543	1414	1119	1242
	NV:	20	19	13	11	13
	CPU:	260.6	556.3	388.6	304.5	200.3
Insert 1 Initial Solution	TT:	<u>3714</u>	<u>3562</u>	<u>3019</u>	<u>2475</u>	<u>2493</u>
	TD:	1872	1704	1533	1248	1343
	NV:	21	19	16	12	12
	CPU:	1.166	1.259	1.462	1.804	1.547
Insert 1 Improved Solution	TT:	[3572]	[3264]	[2642]	2233	2306
	TD:	1672	1531	1370	1183	1313
	NV:	20	19	15	12	12
	CPU:	156.6	257.7	382.5	330.7	288.1
Insert 2 Initial Solution	TT:	4097	3963	3397	2656	2729
	TD:	2231	2214	1725	1423	1595
	NV:	23	22	18	13	13
	CPU:	1.618	1.773	2.145	2.855	2.284
Insert 2 Improved Solution	TT:	3568	3300	2642	2249	2351
	TD:	1679	1559	1407	1179	1268
	NV:	20	19	15	12	13
	CPU:	332.5	642.6	579.6	561.7	473.9
Insert 3 Initial Solution	TT:	4114	3877	3364	3130	3133
	TD:	2246	2100	1983	2079	1959
	NV:	23	21	16	14	14
	CPU:	1.523	1.686	1.869	2.081	1.885
Insert 3 Improved Solution	TT:	3585	3315	2700	2160	[2318]
	TD:	1699	1585	1359	1143	1240
	NV:	20	19	15	11	13
	CPU:	407.2	271.8	465.7	659.5	347.9

Legend: TT: Total Route Time
TD: Total Route Distance
NV: Number of Vehicles
CPU: CPU Seconds UNIVAC 1100/82

Best initial solution underlined
Best final solution boxed

Table 4. Computational Results for the Multiple Vehicle Routing and Scheduling Problem with Time Windows: Test Problems S6 - S10

Problem		S6	S7	S8	S9	S10
Customers		100	100	100	100	100
% Windows		75	90	75	90	75
Best Known Solution	TT:	2744	2540	2230	2524	2118
	TD:	1475	1447	1137	1355	1106
	NV:	14	12	10	13	10
	CPU:	2.9	3.0	3.5	2.9	3.4
Nearest Neighbor Initial Solution	TT:	3166	3134	2877	2934	2637
	TD:	1807	1724	1478	1725	1591
	NV:	15	14	13	14	12
	CPU:	.8630	.8702	.8694	.8540	.8650
Nearest Neighbor Improved Solution	TT:	2494	2375	2127	2428	2128
	TD:	1372	1242	1115	1283	1107
	NV:	14	13	11	14	11
	CPU:	514.3	563.5	428.5	420.3	208.1
Insert 1 Initial Solution	TT:	2724 (underlined)	2493 (underlined)	2320 (underlined)	2644 (underlined)	2325 (underlined)
	TD:	1456	1343	1211	1431	1283
	NV:	14	12	11	14	11
	CPU:	1.421	1.561	1.856	1.418	1.923
Insert 1 Improved Solution	TT:	2464	[2306] (boxed)	[2041] (boxed)	[2351] (boxed)	2132
	TD:	1362	1213	1035	1258	1114
	NV:	14	12	10	13	11
	CPU:	300.6	359.1	514.4	343.6	371.6
Insert 2 Initial Solution	TT:	3007	2729	2514	2817	2419
	TD:	1732	1595	1369	1721	1337
	NV:	15	13	12	14	12
	CPU:	1.936	2.211	3.001	1.971	3.140
Insert 2 Improved Solution	TT:	2459	2351	2100	2476	[2099] (boxed)
	TD:	1346	1268	1096	1315	1057
	NV:	14	13	11	14	11
	CPU:	336.3	477.9	374.6	348.6	274.7
Insert 3 Initial Solution	TT:	3206	3133	2826	3105	3144
	TD:	1968	1959	1818	1763	1932
	NV:	15	14	13	15	14
	CPU:	1.687	1.899	2.222	1.727	2.253
Insert 3 Improved Solution	TT:	[2430] (boxed)	2318	2069	2526	2101
	TD:	1338	1240	1069	1314	1068
	NV:	13	13	11	14	11
	CPU:	349.6	349.5	700.6	292.9	459.3

Legend: TT: Total Route Time
TD: Total Route Distance
NV: Number of Vehicles
CPU: CPU Seconds UNIVAC 1100/82

Best initial solution underlined
Best final solution boxed

Table 5. Computational Results for the Multiple Vehicle Routing and Scheduling Problem with Time Windows: Test Problems S11 - S15

Problem		S11	S12	S13	S14	S15
Customers		100	100	100	100	100
% Windows		75	90	75	90	75
Best Known Solution	TT:	2436	2425	9828	9828	9828
	TD:	1284	1231	828	828	828
	NV:	12	12	10	10	10
	CPU:	3.1	3.1			
Nearest Neighbor Initial Solution	TT:	2970	3072	12377	10812	9872
	TD:	1809	1744	2365	1485	855
	NV:	14	14	12	11	10
	CPU:	.8608	.8682	.8510	.8462	.8454
Nearest Neighbor Improved Solution	TT:	2305	2307	9938	10177	9828
	TD:	1233	1224	938	992	828
	NV:	12	12	11	11	10
	CPU:	535.3	329.2	358.5	472.7	92.01
Insert 1 Initial Solution	TT:	2445	2442	9981	9909	9983
	TD:	1376	1309	976	865	923
	NV:	12	12	10	10	10
	CPU:	1.586	1.549	1.539	1.357	1.324
Insert 1 Improved Solution	TT:	2334	2228	9828	9828	9828
	TD:	1245	1171	828	828	828
	NV:	12	11	10	10	10
	CPU:	212.3	260.9	130.1	118.6	118.8
Insert 2 Initial Solution	TT:	2653	2733	10513	10749	11178
	TD:	1498	1574	1513	1298	1171
	NV:	13	13	11	11	12
	CPU:	2.445	2.323	2.166	1.927	1.815
Insert 2 Improved Solution	TT:	2296	2342	9909	9935	9828
	TD:	1232	1189	909	883	828
	NV:	13	12	11	11	10
	CPU:	284.6	373.6	800.7	367.5	164.8
Insert 3 Initial Solution	TT:	3228	3533	11608	11206	11289
	TD:	1975	2053	2322	2021	1642
	NV:	15	16	11	12	13
	CPU:	1.894	1.864	1.874	1.752	1.653
Insert 3 Improved Solution	TT:	2328	2283	10008	10330	9828
	TD:	1246	1213	997	1189	828
	NV:	13	12	11	11	10
	CPU:	254.5	509.3	238.4	325.6	153.2

Legend: TT: Total Route Time
TD: Total Route Distance
NV: Number of Vehicles
CPU: CPU Seconds UNIVAC 1100/82

Best initial solution underlined
Best final solution boxed

route time and distance among all solutions presented in Solomon's original paper (1983a). The "best" solution obtained for problems S1 - S12 was typically produced by the procedure described as Insert 1 in Section 2.1. For the clustered problems, S13 - S15, the optimal solution is known and is indicated in Table 5. It is noted that the total schedule time is the sum of the total distance traveled, the total waiting time, and the customer service times. For problems S1 - S12, each customer required ten time units to be serviced. In problems S13 - S15, each customer required 90 time units to be serviced.

A third VRSPTW was made available to the authors by Professor Bruce Golden of the University of Maryland. *This real world problem consisted of a single depot and 250 customers with very wide (10 - 12 hour) time windows*. This data set represented one day's deliveries for an institutional food distributor located in the mid-Atlantic region of the United States. In this problem, vehicle capacity was not a consideration, however, customer service time varied across customers. The travel time between customers was computed as a piecewise linear function that allowed average vehicle speed to vary as a function of distance.

Initial application of the Insert 1 route construction procedure to this real world problem produced four vehicle routes. The largest route contained over 90 customers. The subsequent application of the within route 2-opt and 3-opt procedures required several hours of CPU time to complete. As a result, the route construction procedure was reapplied to the original problem and the number of customers per route was restricted to 50. (The initial solution supplied by Professor Golden used 50 customers per route.) Again, the improvement procedures were applied in the manner proposed above and a final improved solution was obtained in a total of 10,000 CPU seconds on a UNIVAC 1100/82 mainframe.

The results produced by the solution improvement procedures on this large scale real world problem were consistent with the results produced for the VRSPTW test problems. A 15.23 percent improvement of an equally weighted sum of route time and route distance was obtained by applying the branch exchange procedures to the time window feasible solution constructed by the Insert 1 procedure. The first application of the within route 2-opt procedure produced a 10.18 percent improvement. The subsequent application of the within route 3-opt procedure produced a 2.57 percent improvement. A 1.59 percent improvement was obtained by applying the between route improvement procedures. The remaining .89 percent was obtained on the second application of the within route improvement algorithms.

4.0 Summary and Conclusions.

A summary of the computational results of Tables 1 through 5 is presented in Table 6. Table 6 presents the average performance of the solution improvement procedures across the three problem types for each of the route construction procedures investigated in this study. The average performance measures presented are the percent reduction in total route time (TT%), the percent reduction in total route distance (TD%), the percent reduction in the number of vehicles required (NV%), and the average processing time in seconds required (CPU) by the improvement procedures. The percent improvements are measured with respect to the initial route determined by each of the route construction procedures. A final measure of performance included for each case in Table 6 is the average percent above the optimal solution (OP%) of the final improved heuristic solution. In the cases of problems S1 - S12, where only a best known solution was known, the negative percentages imply that the improved solution obtained in this study was better than the best previously known solution value. The three problem types presented in Table 6 are the time window constrained traveling salesman problems, B11 - B52, the multiple vehicle routing and scheduling problems without special structure, S1 - S12, and the clustered multiple vehicle problems, S13 - S15.

An examination of the results presented in Tables 1 through 6 indicates that the performance of the route improvement procedures was generally very good. For the structured problems, B11 - B52 and S13 - S15, the optimal solution was identified in 43 of the 52 cases. For the time constrained traveling salesman

TABLE 6. Summary Percent Improvement Figures

Route Construction Procedure		Problems B11 - B52	Problems S1 - S12	Problems S13 - S15
Nearest Neighbor	TT%:	15.0	22.1	8.5
	TD%:	12.6	22.0	11.0
	NV%:	0.0 *	10.5	4.8
	CPU:	162.3	392.5	307.7
	OP%:	0.0 *	-7.57	1.23
Insert 1	TT%:	15.4	8.6	3.4
	TD%:	12.2	10.2	6.9
	NV%:	15.0	2.9	2.8
	CPU:	65.92	367.9	122.5
	OP%:	.16	-8.62	0.0 *
Insert 2	TT%:	17.3	15.2	8.8
	TD%:	12.5	21.7	17.9
	NV%:	5.05	6.8	10.1
	CPU:	118.3	421.7	444.3
	OP%:	.79	-7.52	1.0
Insert 3	TT%:	7.4	24.7	16.1
	TD%:	6.9	35.0	30.2
	NV%:	0.0 *	13.3	16.3
	CPU:	178.2	422.3	239.1
	OP%:	0.0 *	-8.12	5.9

Legend:

TT%: Average percent reduction in total route time
TD%: Average percent reduction in total route distance
NV%: Average percent reduction in number of vehicles
CPU: Average CPU seconds UNIVAC 1100/82
OP%: Average percent above known optimal solution or, when negative, percent below previous best solution

* - indicates optimal value was found in all cases

problems, B11 - B52, the route improvement procedures applied to the initial solutions determined by the nearest neighbor route construction heuristic identified the optimal solution for each of the ten problem instances. This result was not surprising since the time window constrained traveling salesman test problems were generated using a nearest neighbor solution. The Insert 3 procedure also produced the optimal solution for each of the problems B11 - B52 and produced a better initial solution than the nearest neighbor heuristic in all but one of the ten cases. The incorporation of the urgency of visiting each node was obviously an important factor in solving these traveling salesman problems with very tight time window constraints. In the multiple vehicle clustered problems, S13 - S15, the first insertion procedure coupled with the route improvement procedures produced the optimal solution in all three problem instances.

Using an equally weighted combination of time and distance as the measure of performance in the unstructured problems, S1 - S12, the best known solution was improved upon in every case. The average overall improvement with respect to the best known solution for these problems was 7.96 percent. Although the total improvement produced in these tests is a function of both the route construction and the route improvement procedures, the best overall solutions are usually obtained from the best starting solutions, though the percentage improvements may be smaller. Generally the insertion procedures outperformed the nearest neighbor solutions for the unstructured problems. The first insertion criterion yielded the least cost initial solution in all cases and produced the lowest cost final solution in seven of the twelve cases.

In analyzing the effectiveness of the within route and the between route improvement procedures, it was found that generally both the 2-opt and the 3-opt procedures contributed to the improvements. (Details of the average percentage improvements for each of the procedures across test problem type are given in Appendix C.) The 3-opt procedure between two routes, however, was essential for allowing the number of vehicles to be reduced. Overall a 13.6 percent reduction in the number of vehicles required was obtained.

Sketches of the initial solution produced by insertion procedure one and the final solution generated by the solution improvement procedures for problem S1 are presented in Figures 1A and 1B. Similar sketches for problem S15 are presented in Figures 2A and 2B. (The details of these problems are presented in Appendix B. The X and Y axes of the sketches correspond to the (X,Y) coordinates of the test problems.) The improved solutions may be characterized as somewhat "cleaner" (i.e. they have fewer overlapping routes) and more compact than the initial solutions. This is particularly true for the results for problem S15.

Although the solution improvement procedures of this study were effective in reducing the time, distance, and number of vehicles required on the routes, the processing time required to produce these results, especially for the large scale real world problem, was large. This is due, to some extent, to the basic implementation of the algorithms of this study. Additional improvements in efficiency could be made, however, by limiting the branch exchanges of the procedures to only those reconfigurations which preserve the orientation of the original route. In this study, less than 10 percent of the solution improvements involved the reversal of the orientation of a sequence of two or more customers. The effect of such a restriction upon solution quality is a matter for future research. It is noted, however, that a similar procedure proposed by Or (1976) significantly improved the efficiency of the improvement procedures while only moderately degrading solution quality.

ACKNOWLEDGMENTS

The authors wish to express their appreciation to Professor Marius Solomon (Northeastern University) and Professor Bruce Golden (University of Maryland) for providing test problems used in this study.

Interested readers may obtain complete copies of the test

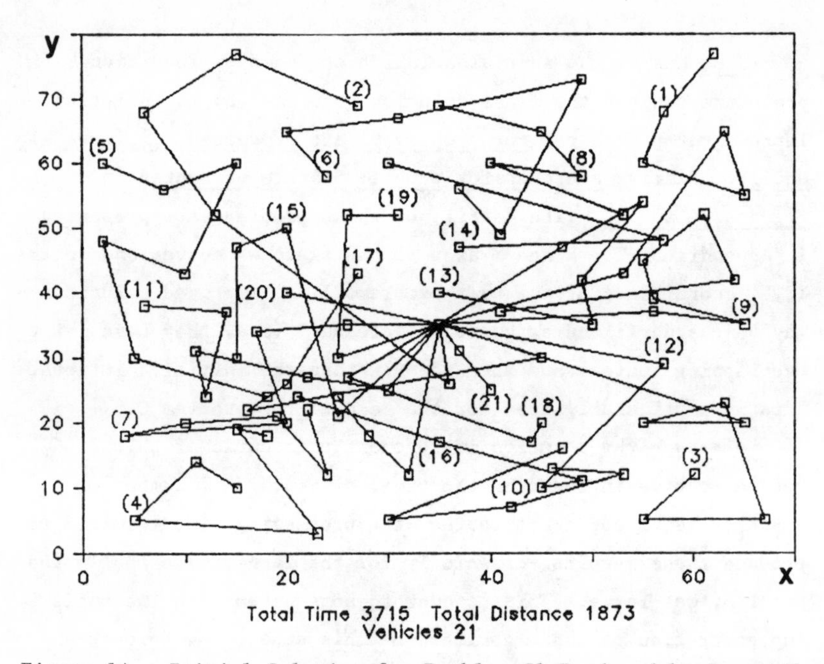

Figure 1A. Initial Solution for Problem S1 Produced by Insert 1

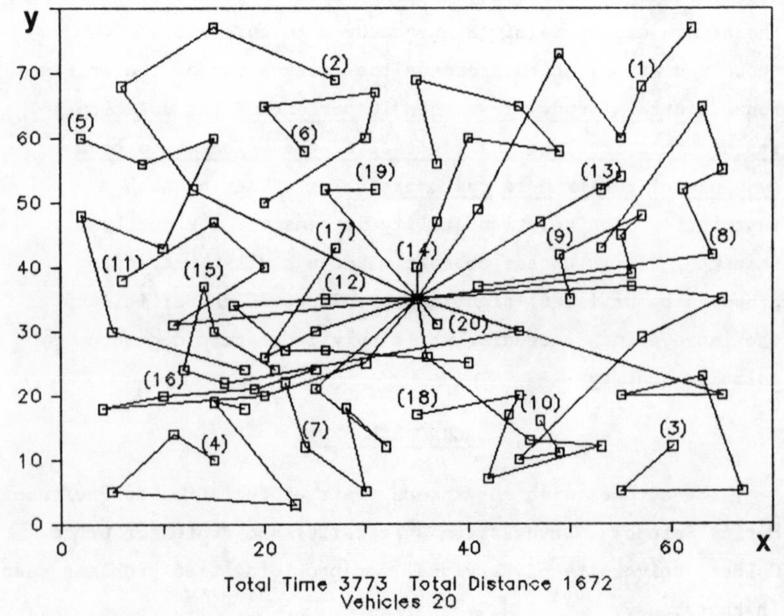

Figure 1B. Final Improved Solution for Problem S1

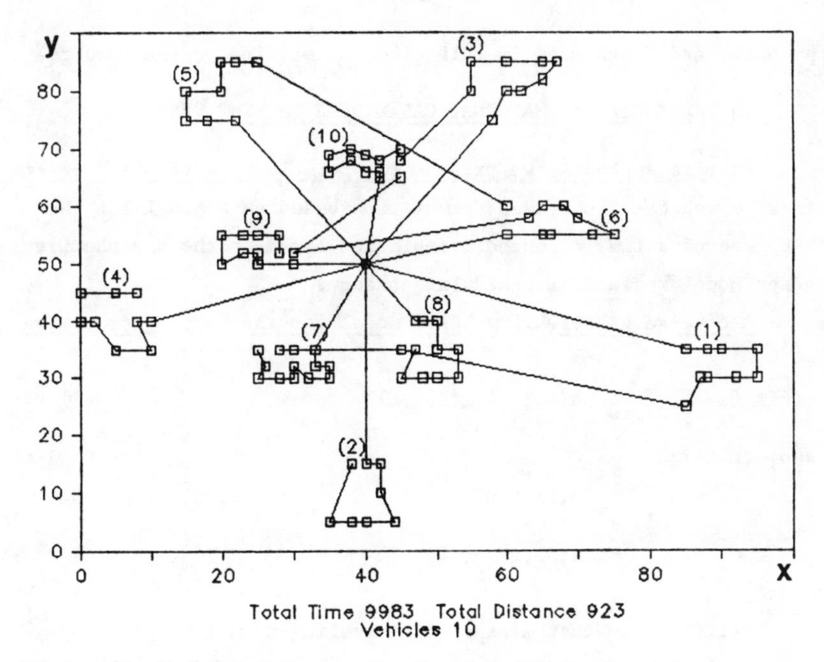

Figure 2A. Initial Solution for Problem S15 Produced by Insert 1

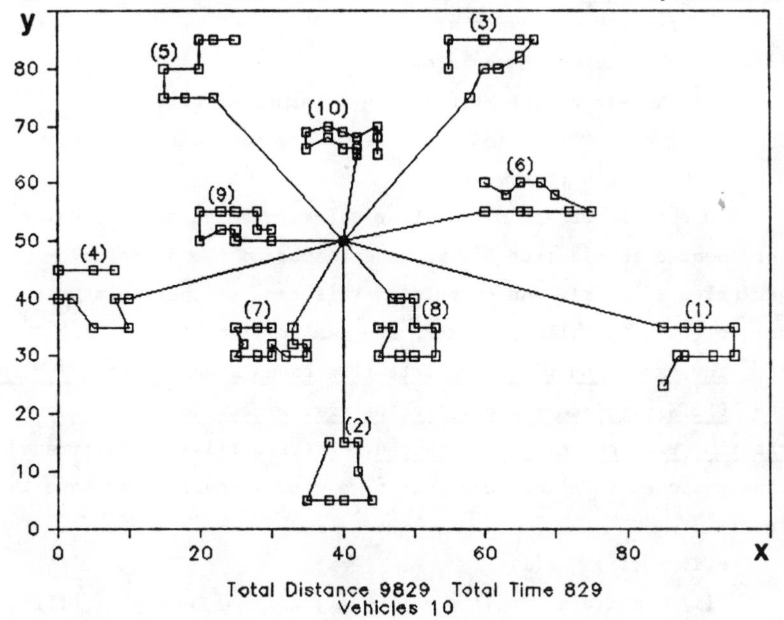

Figure 2B. Final Improved Solution for Problem S15

problems and codes used in this study by writing to the authors.

APPENDIX A: MATHEMATICAL MODELS FOR THE VRSPTW

In this appendix, we present two formulations of the VRSPTW: first a vehicle flow based formulation based on a model for the VRP, second a time oriented formulation based on the disjunctive graph model known from scheduling theory.

A concise vehicle flow based model for the VRP may be stated as:

$$\text{minimize} \quad \sum_{i} \sum_{j} \sum_{k} c(i,j) * x(i,j,k) \qquad (1)$$

subject to:

$$\sum_{i} \sum_{j} u(i) * x(i,j,k) \leq Q \qquad k = 1,2,\ldots,M \qquad (2)$$

$$x(i,j,k) \in S \qquad (3)$$

where:

$c(i,j)$ = the cost measure of traveling from i to j

$u(i)$ = the demand of customer i

Q = the vehicle capacity

M = the number of vehicles

S = the set of all M-traveling salesmen solutions

$$x(i,j,k) = \begin{cases} 1 & \text{if } i \text{ and } j \text{ are served sequentially on route } k \\ 0 & \text{otherwise.} \end{cases}$$

The set, S, of all M-traveling salesmen solutions may be represented as a set of linear constraints on the decision variables $x(i,j,k)$. An example of this type of constraint set may be found in Miller, Tucker, and Zemlin (1960).

The extension of the vehicle flow based model for the VRP to the VRSPTW requires the addition of time window constraints on the time each customer is serviced. Letting $t(i)$ be the time that customer i is serviced, the time window constraints have the form:

$$e(i) \leq t(i) \leq f(i) \qquad (4)$$

$$t(i) = \max(e(i),\ t(j) + s(j) + a(j,i)) \qquad (5)$$

for j the immediate predecessor of i

where:

e(i) = the earliest time that customer i can be serviced

f(i) = the latest time that customer i can be serviced

a(i,j) = the travel time from i to j

s(i) = the service or dwell time at customer i.

The relationship between the customer visitation times, t(i), and the variables x(i,j,k) may be specified through the use of both linear and nonlinear constraints. Explicit examples of these constraints may be found in the work of Sexton (1979) or in Solomon (1983a).

Extending an earlier model proposed for the time window constrained traveling salesman problem, _a second, time oriented, model for the VRSPTW may be proposed_. The time oriented model requires that the travel times of the problem be symmetric and satisfy the triangle inequality. All routes begin at the depot, customer 1, at time t(1) = 0.0. Additionally, the model, as proposed here, seeks to minimize the maximum route length, t(n+1), over all routes considered.

The model may then be stated as:

$$\text{Minimize } t(n+1) - t(1) \tag{6}$$

subject to:

$$t(i) - t(1) \geq a(1,i) \qquad i = 2,3,\ldots,n \tag{7}$$

$$|t(i) - t(j)| \geq a(i,j) * y(i,k) * y(j,k) \tag{8}$$

$$i = 2,3,\ldots,n-1, \quad j = i+1,\ldots,n$$

$$t(n+1) - t(i) \geq a(i,1) \qquad i = 2,3,\ldots,n \tag{9}$$

$$\sum_i u(i) * y(i,k) \leq Q \qquad k = 1,2,\ldots,M \tag{10}$$

$$e(i) \leq t(i) \leq f(i) \qquad i = 1,2,\ldots,n \tag{11}$$

$$\sum_k y(i,k) = 1 \qquad i = 1,2,\ldots,n \tag{12}$$

$$t(i) \geq 0 \qquad i = 1,2,\ldots,n \tag{13}$$

where:

n = the number of customers

t(n+1) = time that the last vehicle returns to the depot

$$y(i,k) = \begin{cases} 1 & \text{if customer } i \text{ is serviced by vehicle } k \\ 0 & \text{otherwise.} \end{cases}$$

A customer may have zero, one, or several time window constraints associated with its delivery time. The problems considered in this study typically have customers with either one or no time window. All of the algorithms considered, however, are easily adapted to the case of multiple time windows for each customer.

APPENDIX B: SAMPLE TEST PROBLEMS AND CODES

In this appendix, two sample test problems and the FORTRAN code of two of the heuristic algorithms used in this study are presented. The test problems shown in detail correspond to S1 and S15. The solutions to these problems are illustrated in Figures 1 and 2. Each data set includes 101 customers. The depot is designated as customer one (1). Each customer is described by several attributes: customer number (N), X coordinate (X), Y coordinate (Y), earliest time window (E), latest time window (F), quantity demanded (Q), and customer service time (S). The quantity demanded for the depot is equal to the vehicle capacity.

DETAIL OF TEST PROBLEM S1

(N)	(X)	(Y)	(E)	(F)	(Q)	(S)
1	35.00	35.00	0.00	0.00	230.00	0.00
2	41.00	49.00	10.00	161.00	171.00	10.00
3	35.00	17.00	7.00	50.00	60.00	10.00
4	55.00	45.00	13.00	116.00	126.00	10.00
5	55.00	20.00	19.00	149.00	159.00	10.00
6	15.00	30.00	26.00	34.00	44.00	10.00
7	25.00	30.00	3.00	99.00	109.00	10.00
8	20.00	50.00	5.00	81.00	91.00	10.00
9	10.00	43.00	9.00	95.00	105.00	10.00
10	55.00	60.00	16.00	97.00	107.00	10.00
11	30.00	60.00	16.00	124.00	134.00	10.00
12	20.00	65.00	12.00	67.00	77.00	10.00
13	50.00	35.00	19.00	63.00	73.00	10.00
14	30.00	25.00	23.00	159.00	169.00	10.00
15	15.00	10.00	20.00	32.00	42.00	10.00
16	30.00	5.00	8.00	61.00	71.00	10.00
17	10.00	20.00	19.00	75.00	85.00	10.00
18	5.00	30.00	2.00	157.00	167.00	10.00
19	20.00	40.00	12.00	87.00	97.00	10.00
20	15.00	60.00	17.00	76.00	86.00	10.00
21	45.00	65.00	9.00	126.00	136.00	10.00
22	45.00	20.00	11.00	62.00	72.00	10.00
23	45.00	10.00	18.00	97.00	107.00	10.00

24	55.00	5.00	29.00	68.00	78.00	10.00
25	65.00	35.00	3.00	153.00	163.00	10.00
26	65.00	20.00	6.00	172.00	182.00	10.00
27	45.00	30.00	17.00	132.00	142.00	10.00
28	35.00	40.00	16.00	37.00	47.00	10.00
29	41.00	37.00	16.00	39.00	49.00	10.00
30	64.00	42.00	9.00	63.00	73.00	10.00
31	40.00	60.00	21.00	71.00	81.00	10.00
32	31.00	52.00	27.00	50.00	60.00	10.00
33	35.00	69.00	23.00	141.00	151.00	10.00
34	53.00	52.00	11.00	37.00	47.00	10.00
35	65.00	55.00	14.00	117.00	127.00	10.00
36	63.00	65.00	8.00	143.00	153.00	10.00
37	2.00	60.00	5.00	41.00	51.00	10.00
38	20.00	20.00	8.00	134.00	144.00	10.00
39	5.00	5.00	16.00	83.00	93.00	10.00
40	60.00	12.00	31.00	44.00	54.00	10.00
41	40.00	25.00	9.00	85.00	95.00	10.00
42	42.00	7.00	5.00	97.00	107.00	10.00
43	24.00	12.00	5.00	31.00	41.00	10.00
44	23.00	3.00	7.00	132.00	142.00	10.00
45	11.00	14.00	18.00	69.00	79.00	10.00
46	6.00	38.00	16.00	32.00	42.00	10.00
47	2.00	48.00	1.00	117.00	127.00	10.00
48	8.00	56.00	27.00	51.00	61.00	10.00
49	13.00	52.00	36.00	165.00	175.00	10.00
50	6.00	68.00	30.00	108.00	118.00	10.00
51	47.00	47.00	13.00	124.00	134.00	10.00
52	49.00	58.00	10.00	88.00	98.00	10.00
53	27.00	43.00	9.00	52.00	62.00	10.00
54	37.00	31.00	14.00	95.00	105.00	10.00
55	57.00	29.00	18.00	140.00	150.00	10.00
56	63.00	23.00	2.00	136.00	146.00	10.00
57	53.00	12.00	6.00	130.00	140.00	10.00
58	32.00	12.00	7.00	101.00	111.00	10.00
59	36.00	26.00	18.00	200.00	210.00	10.00
60	21.00	24.00	28.00	18.00	28.00	10.00
61	17.00	34.00	3.00	162.00	172.00	10.00
62	12.00	24.00	13.00	76.00	86.00	10.00
63	24.00	58.00	19.00	58.00	68.00	10.00
64	27.00	69.00	10.00	34.00	44.00	10.00
65	15.00	77.00	9.00	73.00	83.00	10.00
66	62.00	77.00	20.00	51.00	61.00	10.00
67	49.00	73.00	25.00	127.00	137.00	10.00
68	67.00	5.00	25.00	83.00	93.00	10.00
69	56.00	39.00	36.00	142.00	152.00	10.00
70	37.00	47.00	6.00	50.00	60.00	10.00
71	37.00	56.00	5.00	182.00	192.00	10.00
72	57.00	68.00	15.00	77.00	87.00	10.00
73	47.00	16.00	25.00	35.00	45.00	10.00
74	44.00	17.00	9.00	78.00	88.00	10.00
75	46.00	13.00	8.00	149.00	159.00	10.00
76	49.00	11.00	18.00	69.00	79.00	10.00
77	49.00	42.00	13.00	73.00	83.00	10.00
78	53.00	43.00	14.00	179.00	189.00	10.00
79	61.00	52.00	3.00	96.00	106.00	10.00
80	57.00	48.00	23.00	92.00	102.00	10.00
81	56.00	37.00	6.00	182.00	192.00	10.00
82	55.00	54.00	26.00	94.00	104.00	10.00
83	15.00	47.00	16.00	55.00	65.00	10.00
84	14.00	37.00	11.00	44.00	54.00	10.00
85	11.00	31.00	7.00	101.00	111.00	10.00
86	16.00	22.00	41.00	91.00	101.00	10.00
87	4.00	18.00	35.00	94.00	104.00	10.00
88	28.00	18.00	26.00	93.00	103.00	10.00
89	26.00	52.00	9.00	74.00	84.00	10.00
90	26.00	35.00	15.00	176.00	186.00	10.00

91	31.00	67.00	3.00	95.00	105.00	10.00
92	15.00	19.00	1.00	160.00	170.00	10.00
93	22.00	22.00	2.00	18.00	28.00	10.00
94	18.00	24.00	22.00	188.00	198.00	10.00
95	26.00	27.00	27.00	100.00	110.00	10.00
96	25.00	24.00	20.00	39.00	49.00	10.00
97	22.00	27.00	11.00	135.00	145.00	10.00
98	25.00	21.00	12.00	133.00	143.00	10.00
99	19.00	21.00	10.00	58.00	68.00	10.00
100	20.00	26.00	9.00	83.00	93.00	10.00
101	18.00	18.00	17.00	185.00	195.00	10.00

DETAIL OF TEST PROBLEM S15

(N)	(X)	(Y)	(E)	(F)	(Q)	(S)
1	40.00	50.00	0.00	0.00	1236.00	0.00
2	45.00	68.00	10.00	912.00	967.00	90.00
3	45.00	70.00	30.00	825.00	870.00	90.00
4	42.00	66.00	10.00	65.00	146.00	90.00
5	42.00	68.00	10.00	727.00	782.00	90.00
6	42.00	65.00	10.00	15.00	67.00	90.00
7	40.00	69.00	20.00	621.00	702.00	90.00
8	40.00	66.00	20.00	170.00	225.00	90.00
9	38.00	68.00	20.00	255.00	324.00	90.00
10	38.00	70.00	10.00	534.00	605.00	90.00
11	35.00	66.00	10.00	357.00	410.00	90.00
12	35.00	69.00	10.00	448.00	505.00	90.00
13	25.00	85.00	20.00	652.00	721.00	90.00
14	22.00	75.00	30.00	30.00	92.00	90.00
15	22.00	85.00	10.00	567.00	620.00	90.00
16	20.00	80.00	40.00	384.00	429.00	90.00
17	20.00	85.00	40.00	475.00	528.00	90.00
18	18.00	75.00	20.00	99.00	148.00	90.00
19	15.00	75.00	20.00	179.00	254.00	90.00
20	15.00	80.00	10.00	278.00	345.00	90.00
21	30.00	50.00	10.00	10.00	73.00	90.00
22	30.00	52.00	20.00	914.00	965.00	90.00
23	28.00	52.00	20.00	812.00	883.00	90.00
24	28.00	55.00	10.00	732.00	777.00	90.00
25	25.00	50.00	10.00	65.00	144.00	90.00
26	25.00	52.00	40.00	169.00	224.00	90.00
27	25.00	55.00	10.00	622.00	701.00	90.00
28	23.00	52.00	10.00	261.00	316.00	90.00
29	23.00	55.00	20.00	546.00	593.00	90.00
30	20.00	50.00	10.00	358.00	405.00	90.00
31	20.00	55.00	10.00	449.00	504.00	90.00
32	10.00	35.00	20.00	200.00	237.00	90.00
33	10.00	40.00	30.00	31.00	100.00	90.00
34	8.00	40.00	40.00	87.00	158.00	90.00
35	8.00	45.00	20.00	751.00	816.00	90.00
36	5.00	35.00	10.00	283.00	344.00	90.00
37	5.00	45.00	10.00	665.00	716.00	90.00
38	2.00	40.00	20.00	383.00	434.00	90.00
39	0.00	40.00	30.00	479.00	522.00	90.00
40	0.00	45.00	20.00	567.00	624.00	90.00
41	35.00	30.00	10.00	264.00	321.00	90.00
42	35.00	32.00	10.00	166.00	235.00	90.00
43	33.00	32.00	20.00	68.00	149.00	90.00
44	33.00	35.00	10.00	16.00	80.00	90.00
45	32.00	30.00	10.00	359.00	412.00	90.00
46	30.00	30.00	10.00	541.00	600.00	90.00
47	30.00	32.00	30.00	448.00	509.00	90.00
48	30.00	35.00	10.00	1054.00	1127.00	90.00
49	28.00	30.00	10.00	632.00	693.00	90.00
50	28.00	35.00	10.00	1001.00	1066.00	90.00
51	26.00	32.00	10.00	815.00	880.00	90.00

52	25.00	30.00	10.00	725.00	786.00	90.00
53	25.00	35.00	10.00	912.00	969.00	90.00
54	44.00	5.00	20.00	286.00	347.00	90.00
55	42.00	10.00	40.00	186.00	257.00	90.00
56	42.00	15.00	10.00	95.00	158.00	90.00
57	40.00	5.00	30.00	385.00	436.00	90.00
58	40.00	15.00	40.00	35.00	87.00	90.00
59	38.00	5.00	30.00	471.00	534.00	90.00
60	38.00	15.00	10.00	651.00	740.00	90.00
61	35.00	5.00	20.00	562.00	629.00	90.00
62	50.00	30.00	10.00	531.00	610.00	90.00
63	50.00	35.00	20.00	262.00	317.00	90.00
64	50.00	40.00	50.00	171.00	218.00	90.00
65	48.00	30.00	10.00	632.00	693.00	90.00
66	48.00	40.00	10.00	76.00	129.00	90.00
67	47.00	35.00	10.00	826.00	875.00	90.00
68	47.00	40.00	10.00	12.00	77.00	90.00
69	45.00	30.00	10.00	734.00	777.00	90.00
70	45.00	35.00	10.00	916.00	969.00	90.00
71	95.00	30.00	30.00	387.00	456.00	90.00
72	95.00	35.00	20.00	293.00	360.00	90.00
73	53.00	30.00	10.00	450.00	505.00	90.00
74	92.00	30.00	10.00	478.00	551.00	90.00
75	53.00	35.00	50.00	353.00	412.00	90.00
76	45.00	65.00	20.00	997.00	1068.00	90.00
77	90.00	35.00	10.00	203.00	260.00	90.00
78	88.00	30.00	10.00	574.00	643.00	90.00
79	88.00	35.00	20.00	109.00	170.00	90.00
80	87.00	30.00	10.00	668.00	731.00	90.00
81	85.00	25.00	10.00	769.00	820.00	90.00
82	85.00	35.00	30.00	47.00	124.00	90.00
83	75.00	55.00	20.00	369.00	420.00	90.00
84	72.00	55.00	10.00	265.00	338.00	90.00
85	70.00	58.00	20.00	458.00	523.00	90.00
86	68.00	60.00	30.00	555.00	612.00	90.00
87	66.00	55.00	10.00	173.00	238.00	90.00
88	65.00	55.00	20.00	85.00	144.00	90.00
89	65.00	60.00	30.00	645.00	708.00	90.00
90	63.00	58.00	10.00	737.00	802.00	90.00
91	60.00	55.00	10.00	20.00	84.00	90.00
92	60.00	60.00	10.00	836.00	889.00	90.00
93	67.00	85.00	20.00	368.00	441.00	90.00
94	65.00	85.00	40.00	475.00	518.00	90.00
95	65.00	82.00	10.00	285.00	336.00	90.00
96	62.00	80.00	30.00	196.00	239.00	90.00
97	60.00	80.00	10.00	95.00	156.00	90.00
98	60.00	85.00	30.00	561.00	622.00	90.00
99	58.00	75.00	20.00	30.00	84.00	90.00
100	55.00	80.00	10.00	743.00	820.00	90.00
101	55.00	85.00	20.00	647.00	726.00	90.00

The two algorithms for which program listings are included in this appendix are INSERT1 and the within route 2- and 3-opt procedures. The program INSERT1 is used to read the test problem data set and to create a feasible set of vehicle routes. This feasible solution is then output to an initial solution file. The 2- and 3-opt procedure is then used to attempt to improve these initially constructed solutions. Most of the input and output of the second program is identical to that of INSERT1 and is omitted to save space.

FORTRAN PROGRAM LISTING OF INSERT1

```
      DIMENSION X(150),Y(150),Q(150),E(150),L(150),S(150),A(150)
      DIMENSION D(150,150),V(50,50),QC(50),P(150),IC(150),MM(50)
      DIMENSION IBEST(150),BARR(150)
      DIMENSION IBEST1(150),IBEST2(150),BARR2(150)
      DIMENSION IBEST3(150),BARR3(150)
      INTEGER U,V
      REAL L
      DO 13 I=1,150
      READ(*,99)K,X(I),Y(I),Q(I),E(I),L(I),S(I)
   99 FORMAT(3X,I3,6 (4X,F7.2))
      IF(K.EQ.999)GO TO 17
   13 CONTINUE
   17 N=I-1
      D(1,1)=-1
      DO 30 I=2,N
      DO 30 J=1,I
      IF(I.EQ.J)THEN
      D(I,J)=-1
      ELSE
      D(I,J)=SQRT((X(I)-X(J))**2+(Y(I)-Y(J))**2)
      D(J,I)=D(I,J)
      ENDIF
   30 CONTINUE
      VC=200.
      DATA V/2500*0/,A/150*0./,QC/50*0./,IC/150*0/
      TW=0.
      TS=0.0
      TD=0.0
      A1=1.0
      A2=0.0
      A3=1.0
      A4=2.0
C     PRINT *,A1,A2,A3,A4
      NN=N
      N=N-1
      DO 7 I=1,N
    7 P(I)=I+1
      K=0
   10 M=1
      K=K+1
      V(K,M)=1
C     SELET A FIRST CUSTOMER ON THE ROUTE AS A WEIGHTED
C     COMBINATION OF TIME AND DISTANCE
      CC=1E10
      DO 150 IP=1,N
      J=P(IP)
      IF(QC(K)+Q(J).GT.VC) GO TO 150
      IF(D(1,J).GT.L(J)) GO TO 150
      C=A2*L(J)-A1*D(1,J)
      IF(C.LT.CC) THEN
      CC=C
      IIP=IP
      JJ=J
      ENDIF
  150 CONTINUE
C     CUSTOMER JJ IS CHOSEN TO START VEHICLE ROUTE K
      PRINT *,' CUSTOMER',JJ,' TO START ROUTE',K
      V(K,M+1)=JJ
      A(JJ)=AMAX1(E(JJ),D(1,JJ))
      QC(K)=QC(K)+Q(JJ)
      P(IIP)=P(N)
      N=N-1
      M=M+1
```

```
      MM(K)=2
      IF(N.EQ.0) GO TO 1000
C     USE INSERTION CRITERION 1 TO BUILD THE ROUTE
C     DETERMINE WHERE ON THE ROUTE TO MAKE THE INSERTION
  100 CONTINUE
      V(K,M+1)=1
      CC2=-1E10
      DO 500 IP=1,N
      U=P(IP)
      IF(QC(K)+Q(U).GT.VC) GO TO 500
C     FIND BEST SLOT IN ROUTE FOR CUSTOMER U
      CC1=1E10
      DO 400 MK=1,M
C     CHECK FEASIBILITY OF THE INSERTION
      I=V(K,MK)
      J=V(K,MK+1)
      IF(L(U).LT.A(I)+S(I)+D(I,U)) GO TO 400
      ARRU=AMAX1(E(U),A(I)+S(I)+D(I,U))
      IF(L(J).LT.ARRU+S(U)+D(U,J)) GO TO 400
C     I-U-J IS FEASIBLE CHECK THE PUSH FORWARD
      IF(J.EQ.1) GO TO 420
      LAST=J
      ARRL=AMAX1(E(LAST),ARRU+S(U)+D(U,LAST))
      LM=MK+2
  410 NEXT=V(K,LM)
      ARRN=AMAX1(E(NEXT),ARRL+S(LAST)+D(LAST,NEXT))
      IF(ARRN.GT.L(NEXT)) GO TO 400
      IF(NEXT.EQ.1) GO TO 420
      ARRL=ARRN
      LAST=NEXT
      LM=LM+1
      GO TO 410
  420 CONTINUE
C     U IS FEASIBLE FOR THE ENTIRE ROUTE
C     CALCULATE C1=ADDITIONAL DISTANCE ADDED TO ROUTE
      C1=D(I,U)+D(U,J)-A3*D(I,J)
      IF(C1.GE.CC1) GO TO 400
      CC1=C1
      MMM=MK
C     WRITE(6,466) I,U,J,CC1,MMM
  466 FORMAT(' I,U,J,CC1,MMM',3I5,F7.2,I5)
  400 CONTINUE
      IF(CC1.GT.1E9) GO TO 500
C     A MINIMUM INSERTION POSITION HAS BEEN IDENTIFIED
CC    PRINT *,' MIN INSERT POS',U,' AFTER CUSTOMER',V(K,MMM),'

C1=',CC1
C     CALCULATE POTENTIAL SAVINGS FOR THIS INSERTION
      C2=A4*D(1,U)-CC1
      IF(C2.LE.CC2) GO TO 500
      CC2=C2
      INU=U
      INP=IP
      INS=MMM
C     PRINT *,' CUSTOMER U=',U,' POTENTIAL SAVINGS C2=',C2
  500 CONTINUE
      IF(CC2.LT.-1E9) THEN
      MM(K)=M
      GO TO 10
      ENDIF
C     UPDATE ROUTE TO INCLUDE U
      I1=INU
      DO 600 MK=INS,M
      ITEMP=V(K,MK+1)
      V(K,MK+1)=I1
      LL=V(K,MK)
      A(I1)=AMAX1(E(I1),A(LL)+S(LL)+D(LL,I1))
```

```
      I1=ITEMP
  600 CONTINUE
      QC(K)=QC(K)+Q(INU)
      PRINT *,'CUSTOMER',INU,'ADDED TO VEHICLE',K,'POSITION',MMM
C     WRITE(6,666) (V(K,II),II=1,M+1)
C     WRITE(6,667) (A(V(K,JJ)),JJ=1,M+1)
  666 FORMAT(' CURRENT VEHICLE ROUTE',/10I8)
  667 FORMAT(10F8.2)
      P(INP)=P(N)
      N=N-1
      M=M+1
      MM(K)=M
      IF(N.GT.0) GO TO 100
 1000 CONTINUE
      DO 1011 I=1,K
      PRINT *,'ROUTE ',I
      WRITE(*,991)
      LL=MM(I)
C     CALCULATE THE TOTAL QUANTITY AND WAIT ON THIS ROUTE
      TOTQ=0.0
      TOTW=0.0
      DO 1010 M=1,LL
      J=V(I,M)
      IF(E(J).LE.A(J).AND.L(J).GE.A(J)) THEN
      IC(J)=1
      ELSE
      PRINT *,'ERROR CUST: ',J
      ENDIF
      TOTQ=TOTQ+Q(J)
      WAIT=AMAX1(0.0,E(J)-A(J))
      IF(M.GT.1) THEN
      JM1=V(I,M-1)
      TD=TD+D(JM1,J)
      WAIT=AMAX1(0.0,E(J)-A(JM1)-S(JM1)-D(JM1,J))
      ENDIF
      TOTW=TOTW+WAIT
      WRITE(*,992)V(I,M),A(V(I,M)),WAIT,Q(J)
1010  CONTINUE
      TARR=A(V(I,LL))+D(V(I,LL),1)+S(V(I,LL))
      WRITE(*,992) V(1,1),TARR,TOTW,TOTQ
      TS=TS+TARR
      TD=TD+D(V(I,LL),1)
      TW=TW+TOTW
1011  CONTINUE
      PRINT *,'NUMBER OF ROUTES = ',K
      PRINT *,'TOTAL SCHEDULE TIME =    ',TS
      PRINT *,'TOTAL ROUTE DISTANCE = ',TD
      PRINT *,'TOTAL WAITING TIME =     ',TW
C     DO 1020 I=1,NN
C     IF(IC(I).NE.1) PRINT *,'NOT FOUND: ',I
 1020 CONTINUE
C     WRITE(8,993)((V(I,M),M=1,MM(I)),I=1,K),V(K,M+1)
 991  FORMAT(1X,'CUSTOMER',2X,'ARRIVAL TIME  WAIT  DEMAND')
 992  FORMAT(4X,I3,8X,3F7.2)
 993  FORMAT(20 (1X,I3))
 8000 CONTINUE
C     WRITE SOLUTION FILE TO UNIT 9
      WRITE(9,8008) K
 8008 FORMAT(16I5)
      DO 8010 I=1,K
      WRITE(9,8008) MM(I)
      WRITE(9,8008) (V(I,J),J=1,MM(I))
      WRITE(9,8009) (A(V(I,J)),J=1,MM(I))
 8009 FORMAT(10F8.2)
 8010 CONTINUE
      STOP
      END
```

FORTRAN PROGRAM LISTING OF WITHIN ROUTE 2- AND 3-OPT

```
C      DIMENSION STATEMENTS AS IN INSERT1
C      READ ORIGINAL PROBLEM DATA AS IN INSERT1
C      SET VEHICLE CAPACITY
       VC=230.
C      READ DATA FROM INITIAL SOLUTON FILE
C      THE FOLLOWING CODE IS A 2-OPT PROCEDURE
       IF(ITWO.EQ.1) GO TO 2000
       IF(ITHREE.EQ.1) GO TO 3000
       NUMCHA=0
       NUMREV=0
       DO 1500 KK=1,K
       N=MM(KK)
       V(KK,N+1)=1
 1050  BESTIM=0.0
       BESDIS=0.0
       ARRI=0.0
       DO 1100 II=1,N+1
       IBEST(II)=V(KK,II)
       IF(II.EQ.1) GO TO 1100
       I=V(KK,II-1)
       J=V(KK,II)
       BESDIS=BESDIS+D(I,J)
       BESTIM=AMAX1(E(J),ARRI+S(I)+D(I,J))
       ARRI=BESTIM
 1100  CONTINUE
C      PRINT *,' ROUTE,TIME,DISTANCE',KK,BESTIM,BESDIS
       IF(N.LT.3) GO TO 1500
       DO 1400 II=2,N-1
       DO 1300 III=II+1,N
       DO 1150 JJ=1,N
       IF(JJ.LT.II) THEN
       IBEST(JJ)=V(KK,JJ)
       GO TO 1150
       ENDIF
       IF(JJ.GE.II.AND.JJ.LE.III) THEN
       IPOS=III-JJ+II
       IBEST(JJ)=V(KK,IPOS)
       GO TO 1150
       ENDIF
       IF(JJ.GT.III) THEN
       IBEST(JJ)=V(KK,JJ)
       ENDIF
 1150  CONTINUE
C      PRINT *,'EXCHANGE I AND J',IBEST(II),IBEST(III)
C      CALCULATE TIME AND DISTANCE OF IBEST
C      ALSO CHECK FEASIBILITY OF TIME WINDOWS
       DIS=0.0
       TIME=0.0
       ARRI=0.0
       BARR(1)=0.0
       DO 1200 JJ=2,N+1
       I=IBEST(JJ-1)
       J=IBEST(JJ)
       ARRJ=AMAX1(E(J),ARRI+S(I)+D(I,J))
       IF(ARRJ.GT.L(J)) THEN
       IBEST(II)=V(KK,II)
       IBEST(III)=V(KK,III)
       GO TO 1300
       ENDIF
       ARRI=ARRJ
       DIS=DIS+D(I,J)
       TIME=ARRJ
       BARR(JJ)=ARRJ
 1200  CONTINUE
C      PRINT *,'NEW ROUTE FEAS.,ROUTE,TIME,DISTANCE',KK,TIME,DIS
```

```
C      WRITE(6,1266) (IBEST(LL),LL=1,N+1)
C      WRITE(6,1267) (BARR(LL),LL=1,N+1)
 1266 FORMAT('FEASIBLE ROUTE',/10I8)
 1267 FORMAT(10F8.2)
      A31=.5
      A32=.5
      C3=A31*TIME+A32*DIS
      CC3=A31*BESTIM+A32*BESDIS
      IF(C3.GE.CC3) THEN
      DO 1280 LL=1,N
      IBEST(LL)=V(KK,LL)
 1280 CONTINUE
      GO TO 1300
      ELSE
      PRINT *,'A 2-OPT COST REDUCTION FOUND,C3,CC3',C3,CC3
      PRINT *,'II = ',V(KK,II),' III = ',V(KK,III)
      NUMCHA=NUMCHA+1
      IF(III-II.GT.1) NUMREV=NUMREV+1
      DO 1290 LL=2,N
      V(KK,LL)=IBEST(LL)
      A(V(KK,LL))=BARR(LL)
 1290 CONTINUE
      GO TO 1050
      ENDIF
 1300 CONTINUE
 1400 CONTINUE
 1500 CONTINUE
      ITWO=1
      TS=0.0
      TD=0.0
      TW=0.0
      PRINT *,' WITHIN ROUTE 2-OPT RESULTS'
      GO TO 1000
 2000 CONTINUE
C     GIVEN A 2-OPT SOLUTION THE FOLLOWING CODE
C     PERFORMS A 3-OPT PROCEDURE
      IF(ITHREE.EQ.1) GO TO 3000
      DO 2600 KK=1,K
      N=MM(KK)
      IF(N.LT.4) GO TO 2600
      V(KK,N+1)=1
 2050 BESTIM=0.0
      BESDIS=0.0
      ARRI=0.0
      DO 2100 II=1,N+1
      IBEST(II)=V(KK,II)
      IF(II.EQ.1) GO TO 2100
      I=V(KK,II-1)
      J=V(KK,II)
      BESDIS=BESDIS+D(I,J)
      BESTIM=AMAX1(E(J),ARRI+S(I)+D(I,J))
      ARRI=BESTIM
C     PRINT *,',I,J,BESDIS,BESTIM'I,J,BESDIS,BESTIM
 2100 CONTINUE
C     PRINT *,' ROUTE,TIME,DISTANCE',KK,BESTIM,BESDIS
      DO 2500 II=1,N-2
      DO 2400 III=II+1,N-1
      DO 2300 IIII=III+1,N
      DO 2250 JJJ=1,4
      IF(JJJ.EQ.1) THEN
      DO 2140 LL=1,N
      IBEST(LL)=V(KK,LL)
      IF(LL.GT.II.AND.LL.LE.III) THEN
      IPOS=III-(LL-II-1)
      IBEST(LL)=V(KK,IPOS)
      ENDIF
      IF(LL.GT.III.AND.LL.LE.IIII) THEN
```

```
      IPOS=IIII-(LL-III-1)
      IBEST(LL)=V(KK,IPOS)
      ENDIF
 2140 CONTINUE
      ENDIF
      IF(JJJ.EQ.2) THEN
      DO 2150 LL=1,N
      IBEST(LL)=V(KK,LL)
      IF(LL.GT.II.AND.LL.LE.II+IIII-III) THEN
      IPOS=LL-II+III
      IBEST(LL)=V(KK,IPOS)
      ENDIF
      IF(LL.GT.II+IIII-III.AND.LL.LE.IIII) THEN
      IPOS=LL-IIII+III
      IBEST(LL)=V(KK,IPOS)
      ENDIF
 2150 CONTINUE
      ENDIF
      IF(JJJ.EQ.3) THEN
      DO 2160 LL=1,N
      IBEST(LL)=V(KK,LL)
      IF(LL.GT.II.AND.LL.LE.II+IIII-III) THEN
      IPOS=LL-II+III
      IBEST(LL)=V(KK,IPOS)
      ENDIF
      IF(LL.GT.II+IIII-III.AND.LL.LE.IIII) THEN
      IPOS=IIII-LL+II+1
      IBEST(LL)=V(KK,IPOS)
      ENDIF
 2160 CONTINUE
      ENDIF
      IF(JJJ.EQ.4) THEN
      DO 2170 LL=1,N
      IBEST(LL)=V(KK,LL)
      IF(LL.GT.II.AND.LL.LE.II+IIII-III) THEN
      IPOS=IIII-(LL-II-1)
      IBEST(LL)=V(KK,IPOS)
      ENDIF
      IF(LL.GT.II+IIII-III.AND.LL.LE.IIII) THEN
      IPOS=LL-IIII+III
      IBEST(LL)=V(KK,IPOS)
      ENDIF
 2170 CONTINUE
      ENDIF
C     PRINT *,'EXCHANGE I,J,K',IBEST(II),IBEST(III),IBEST(IIII)
C     CALCULATE TIME AND DISTANCE OF IBEST
C     ALSO CHECK FEASIBILITY OF TIME WINDOWS
      DIS=0.0
      TIME=0.0
      ARRI=0.0
      BARR(1)=0.0
      DO 2200 JJ=2,N+1
      I=IBEST(JJ-1)
      J=IBEST(JJ)
      ARRJ=AMAX1(E(J),ARRI+S(I)+D(I,J))
      IF(ARRJ.GT.L(J)) THEN
      DO 2190 LL=1,N
      IBEST(LL)=V(KK,LL)
 2190 CONTINUE
      GO TO 2250
      ENDIF
      ARRI=ARRJ
      DIS=DIS+D(I,J)
      TIME=ARRJ
      BARR(JJ)=ARRJ
 2200 CONTINUE
C     PRINT *,'NEW ROUTE FEAS.,ROUTE,TIME,DISTANCE',KK,TIME,DIS
```

```
C     WRITE(6,1266) (IBEST(LL),LL=1,N+1)
C     WRITE(6,1267) (BARR(LL),LL=1,N+1)
      A31=.5
      A32=.5
      C3=A31*TIME+A32*DIS
      CC3=A31*BESTIM+A32*BESDIS
      IF(C3.GE.CC3) THEN
      DO 2210 LL=1,N
      IBEST(LL)=V(KK,LL)
 2210 CONTINUE
      GO TO 2250
      ELSE
      PRINT *,'A 3-OPT COST REDUX,C3,CC3,JJJ',C3,CC3,JJJ
      NUMCHA=NUMCHA+1
      IF(JJJ.EQ.1.AND.III-II.GT.1.OR.JJJ.EQ.1.AND.IIII-III.GT.1)
     + NUMREV=NUMREV+1
      IF(JJJ.EQ.3.AND.III-II.GT.1) NUMREV=NUMREV+1
      IF(JJJ.EQ.4.AND.IIII-III.GT.1) NUMREV=NUMREV+1
      DO 2220 II1=2,N
      V(KK,II1)=IBEST(II1)
      A(IBEST(II1))=BARR(II1)
 2220 CONTINUE
C     WRITE(6,1266) (V(KK,LL),LL=1,N+1)
C     WRITE(6,1267) (A(V(KK,LL)),LL=1,N+1)
      GO TO 2050
      ENDIF
 2250 CONTINUE
 2300 CONTINUE
 2400 CONTINUE
 2500 CONTINUE
 2600 CONTINUE
      ITHREE=1
      TS=0.0
      TD=0.0
      TW=0.0
      PRINT *,' WITHIN ROUTE 3-OPT RESULTS'
      PRINT *,'NUMBER IMPROVEMENTS:',NUMCHA,'REVERSALS:',NUMREV
      GO TO 1000
 3000 CONTINUE
 8000 CONTINUE
C     WRITE OUT SOLUTION TO FILE
      END
```

APPENDIX C: DETAIL OF AVERAGE PERCENTAGE IMPROVEMENTS

In this appendix, the average percentage improvements of the solution improvement procedures for the VRSPTW are presented. Table C.1 gives the average percentage improvement by algorithm type for each of the test data sets. The solution value in each case was measured as an equally weighted sum of the total route time and the total route distance. The initial raw percentages were obtained by subtracting the improved solution value from the prior solution and then dividing the result by the original solution value obtained from the route construction procedure.

The abbreviations used in Table C.1 are:

Table C.1. Average Percentage Improvements From Initially Constructed Solutions by Algorithm Type and Across Problem Type

	Problems B11 - B52	Problems S1 - S12	Problems S13 - S15
Nearest Neighbor			
1R 2-opt	6.75	2.83	.84
1R 3-opt	16.02	1.61	.00
Replications	1	3.17	1.67
2R 2-opt	N/A	7.50	4.84
2R 3-opt	N/A	14.13	5.86
3R 3-opt	N/A	.48	.70
Replications	0	2.83	1.67
Total	13.87	26.55	12.02
Insert 1			
1R 2-opt	.32	1.39	.15
1R 3-opt	8.28	.17	.00
Replications	1.22	2.83	1.33
2R 2-opt	.50	2.09	.19
2R 3-opt	13.85	5.92	1.71
3R 3-opt	N/A	.12	N/A
Replications	1	2.42	1
Total	14.05	9.69	2.04
Insert 2			
1R 2-opt	2.65	1.76	.25
1R 3-opt	7.23	0.55	.00
Replications	1.44	3.42	3
2R 2-opt	.67	5.00	3.04
2R 3-opt	11.55	12.76	7.80
3R 3-opt	N/A	.33	.34
Replications	1	3.08	2.67
Total	15.30	20.40	11.32
Insert 3			
1R 2-opt	1.56	4.81	1.28
1R 3-opt	7.90	.43	.51
Replications	1	3.08	1.33
2R 2-opt	N/A	12.73	8.64
2R 3-opt	N/A	13.48	6.46
3R 3-opt	N/A	1.47	.96
Replications	0	2.67	1
Total	7.19	32.87	17.17

i)	1R 2-opt	2-opt within a single route
ii)	1R 3-opt	3-opt within a single route
iii)	2R 2-opt	2-opt between two routes
iv)	2R 3-opt	3-opt between two routes
v)	3R 3-opt	3-opt among three routes.

The average number of replications of each of the within route and the between route procedures is also reported as applied for each initial route construction procedure. The term "N/A" is used to indicate that a particular algorithm was not applicable to a particular instance (the 2R 2-opt procedure, for example, is not applicable to a solution with only one route).

REFERENCES

Baker, E. (1983). An exact algorithm for the time constrained traveling salesman problem. Operations Research, 31, 938-945.

Baker, E.; Bodin, L.; Finnegan, W.; and Ponder, R. (1979). Efficient heuristic solutions to an airline crew scheduling problem. AIIE Transactions, 11, 119-126.

Bell, W.; Dalberto, L.; Fisher, M.; Greenfield, A.; Jaikumar, R.; Mack, R.; and Prutzman, P. (1983). Improving the distribution of industrial gases with an on-line computerized routing and scheduling system. Interfaces, 13, 4-23.

Bodin, L.; Golden, B.; Assad, A.; and Ball, M. (1983). The routing and scheduling of vehicles and crews. Computers and Operations Research, 10, 63-211.

Christofides, N.; Mingozzi, A.; and Toth, P. (1981). State-space relaxation procedures for the computation of bounds to routing problems. Networks, 11, 145-164.

Christofides, N.; Mingozzi, A.; and Toth, P. (1979). The vehicle routing problem. In Combinatorial Optimization, ed. by N. Christofides, A. Mingozzi, P. Toth, and C. Sardi. John Wiley & Sons, Inc. New York, pp.

Desrosiers, A; Soumis, F.; and Desrochers, M. (1983). Routing with time windows by column generation. Working paper 277, Centre de Reserche sur les Transports, University of Montreal, Montreal, Canada.

Eilon, S.; Watson-Gandy, C.; and Christofides, N. (1971). Distribution Management. Griffin Press, London.

Gaskell, T. (1967). Bases for vehicle fleet scheduling. Operational Research Quarterly, 18, 281-295.

Golden, B.; Magnanti, T.; and Nguyen, H. (1977). Implementing vehicle routing algorithms. Networks, 17, 113-148.

Fisher, M.; Greenfield, A.; Jaikumar, R.; and Kedio, P. (October 1982). Real-time scheduling of a bulk delivery fleet: a practical application of Lagrangian relaxation. University of Pennsylvania Decision Sciences Working Paper.

Lenstra, J. and Rinnooy Kan, A. (1981). Complexity of vehicle routing and scheduling problems. Networks, 11, 221-227.

Lin, S. (1965). Computer solutions of the traveling salesman problem. Bell System Technical Journal, 44, 2245-2269.

Lin, S. and Kernighan, B. (1973). An effective heuristic for the traveling salesman problem. Operations Research, 21, 498-516.

Magnanti, T. (1981). Combinatorial optimization and vehicle fleet planning: perspectives and prospects. Networks, 11, 179-216.

Miller, C.; Tucker, A.; and Zemlin, R. (1960). Integer programming formulations and traveling salesman problems. Journal of the ACM, 7, 326-329.

Or, I. (1976). Traveling salesman-type combinatorial problems and their relation to the logistics of regional blood banking. Doctoral Dissertation, Northwestern University.

Psaraftis, H. (1980). A dynamic programming solution to the single vehicle many-to-many immediate request dial-a-ride problem. Transportation Science, 14, 130-154.

Psaraftis, H. (1983). K-interchange procedures for local search in a precedence-constrained routing problem. European Journal of Operational Research, 25, 517-524.

Russell, R. (1977). An effective heuristic for the m-tour traveling salesman problem with some side conditions. Operations Research, 25, 517-524.

Sexton, T. (1979). The single vehicle many-to-many routing and scheduling problem. Ph.D. Thesis, SUNY at Stony Brook.

Solomon, M. (1983a). Vehicle routing and scheduling with time window constraints: models and algorithms. Working Paper 83-42, College of Business Administration, Northeastern University. Forthcoming in Operations Research.

Solomon, M. (1983b). On the worst-case performance of some heuristics for the vehicle routing and scheduling problem with time window constraints. Decision Sciences Working Paper 83-05-03, University of Pennsylvania. Forthcoming in Networks.

Received 2/85; Revised 2/3/86.

AMERICAN JOURNAL OF MATHEMATICAL AND MANAGEMENT SCIENCES

A DYNAMIC PROGRAMMING SOLUTION OF THE LARGE-SCALE SINGLE-VEHICLE DIAL-A-RIDE PROBLEM WITH TIME WINDOWS

Jacques Desrosiers
École des Hautes Études Commerciales de Montréal
Montréal, Canada H3T 1V6

Yvan Dumas
Centre de recherche sur les transports, Université de Montréal
Montréal, Canada H3C 3J7

François Soumis
École Polytechnique de Montréal
Montréal, Canada H3C 3A7

SYNOPTIC ABSTRACT

The single-vehicle dial-a-ride problem with time window constraints for both pick-up and delivery locations, and precedence and capacity constraints, is solved using a forward dynamic programming algorithm. The total distance is minimized. The development of criteria for the elimination of infeasible states results in solution times which increase linearly with problem size.

Key Words and Phrases: dial-a-ride; dynamic programming; routing; scheduling

1986, VOL. 6, NOS. 3 & 4, 301-325
0196-6324/86/030301-25 $8.00

1. INTRODUCTION.

The problem examined in this article is the Single-Vehicle Dial-a-Ride Problem. In this problem, a vehicle picks up riders at their place of origin and takes them to their destination, without exceeding vehicle capacity, and respecting time window constraints at both the origins and the destinations. The objective is to find the itinerary which minimizes the total distance traveled.

1.1. Literature review. The single-vehicle problem has been studied by Sexton and Bodin (1985a, 1985b) who investigated the dial-a-ride problem in which each customer specifies a desired time for pick-up or delivery. The result is a heuristic routing and scheduling algorithm, based on Benders' decomposition, which minimizes total customer inconvenience and is known to produce high quality solutions. Psaraftis (1980) developed a backward dynamic programming method for the single-vehicle problem, but rather than imposing time window constraints at the origins and destinations, he imposed a maximum position shift with the respect to the ordering of pick-up and delivery times requested. Computation time for n customers is $O(n^2 3^n)$. Armstrong and Garfinkel (1982) also developed a backward recursion procedure for this problem using a maximum position shift formulation rather than time windows; their procedure is much more powerful for heavily constrained problems. It allows a great reduction in the number of states generated, but the algorithm cannot optimally solve the case where clock time is used to define window constraints. Psaraftis (1982) developed a forward dynamic programming method with true clock time windows at both pick-up and delivery locations. His objective function minimizes the total time needed to service all customers. Unfortunately, this algorithm is independent of the tightness of the time window constraints, and in practice it cannot solve problems with more than about eight to ten customers. Psaraftis (1983) also analyzes an $O(n^2)$ heuristic for the single-vehicle problem.

Bodin, Golden, Assad and Ball (1983), in their state of the art article, describe three algorithms for the multi-vehicle problem : the NEIGUT/NBS algorithm (a sequential insertion procedure), the Jaw, Odoni, Psaraftis and Wilson (1981) algorithm (a concurrent procedure) and the Bodin and Sexton (1986) algorithm (a clustering/concurrent insertion procedure). Two other algorithms have recently been developped at the University of Montreal : the work of Roy, Rousseau, Lapalme and Ferland (1984a, 1984b), which is also based on a concurrent insertion procedure; and the algorithm of Desrosiers, Dumas and Soumis (1984), a "mini-cluster first, routing second" approach using a column generation scheme.

This paper presents an optimal solution to the single-vehicle problem with clock time windows using a forward dynamic programming approach that significantly reduces the number of states generated. We also use the two dimensional (time, cost) labeling introduced in Desrosiers, Pelletier and Soumis (1983) for solving the shortest path problem with time windows by dynamic programming. We minimize the total distance traveled and not the total time required to serve all customers. The algorithm proposed can efficiently solve problems of up to 40 service requests and is presently used to optimize the routes constructed in a large scale multi-vehicle problem : Bélisle, Soumis, Roy, Desrosiers, Dumas and Rousseau (1984) and Desrosiers, Dumas and Soumis (1984).

1.2. Formulation. Consider a set of n service requests. Associated with each request we have

1) an origin node i with a departure time interval $[A_i, B_i]$, an integer value $C_i \geq 1$ representing the number of people to be picked up and M_i, the time required to handle pick-up;

2) a destination node $n + i$ with an arrival time interval $[A_{n+i}, B_{n+i}]$, a value $C_{n+i} = -C_i$ representing the number of people to be dropped off and M_{n+i}, the drop-off time.

In addition to these 2n nodes, we have a node 0 to represent the vehicle departure point (with time interval $[A_0, B_0]$) and node $2n + 1$ for the vehicle arrival point (with time interval $[A_{2n+1}, B_{2n+1}]$). D_{ij} and T_{ij} respectively denote the distance traveled and the travel time between nodes i and j. The constant C represents the capacity limit of the vehicle.

The formulation includes <u>flow variables</u> x_{ij}, <u>time variables</u> t_i and <u>vehicle load variables</u> y_i, where

$$x_{ij} = \begin{cases} 1 & \text{if the route uses arc } (i,j) \\ 0 & \text{otherwise,} \end{cases}$$

$t_i = \{\text{arrival time at node } i\}$,

$y_i = \{\text{the number of passengers in the vehicle on its departure from node } i\}$.

The shortest route satisfying precedence constraints, time window constraints and vehicle capacity constraints is the solution of the problem :

$$\text{Min } Z = \sum_{i=0}^{2n} \sum_{j=1}^{2n+1} D_{ij} x_{ij} \tag{1}$$

Subject to

$$\sum_{j=1}^{n} x_{0j} = 1 \tag{2}$$

$$\sum_{i=n+1}^{2n} x_{i,2n+1} = 1 \tag{3}$$

$$\sum_{i=0}^{2n} x_{i\ell} = \sum_{j=1}^{2n+1} x_{\ell j} = 1 \qquad \ell = 1,\dots,2n \tag{4}$$

$$x_{ij} \text{ binary} \qquad \begin{matrix} i = 0,1,\dots,2n \\ j = 1,\dots,2n+1 \end{matrix} \tag{5}$$

$$A_i \leq t_i \leq B_i \qquad i = 0,1,\ldots,2n+1 \qquad (6)$$

$$t_i + M_i + T_{i,n+i} \leq t_{n+i} \qquad i = 1,\ldots,n \qquad (7)$$

$$x_{ij} = 1 \Rightarrow t_i + M_i + T_{ij} \leq t_j \qquad \begin{array}{l} i = 0,1,\ldots,2n \\ j = 1,\ldots,2n+1 \end{array} \qquad (8)$$

$$y_0 = 0 \; ; \; 0 \leq y_i \leq C \qquad i = 1,\ldots,2n+1 \qquad (9)$$

$$x_{ij} = 1 \Rightarrow y_i + C_j = y_j \qquad \begin{array}{l} i = 0,1,\ldots,2 \\ j = 1,\ldots,2n+1. \end{array} \qquad (10)$$

This formulation includes a flow problem structure (1-5) where each node is visited only once (4) with a single vehicle departing from node 0 (2) and terminating at node 2n+1 (3). The problem also includes time window constraints (6), origin-destination precedence constraints (7), and vehicle capacity constraints (9). Constraints (8) describe the relation between the flow variables and the time variables, while constraints (10) describe the relation between the flow variables and vehicle load at each node. The problem is essentially a traveling salesman problem with additional constraints, with the non-linear constraints (8) ensuring sub-tour elimination. In fact, these constraints require increasing times at each of the nodes along the route, thus preventing sub-tour formation.

By minimizing the total distance traveled, riding time within the vehicle is not minimized in equation (8). This objective function is less general than others proposed for minimizing user inconvenience. Using the definition of time windows given in Roy, Rousseau, Lapalme and Ferland (1984), it is possible to take user inconvenience issues into account. For example, if a request i has a desired departure time A_i, the latest time the vehicle may arrive at the origin node is defined as $B_i = A_i + a$, with $a \leq 45$ minutes depending on the priority of the request. The lower and upper bounds of the time window at the destination node $n + i$ are given by $A_{n+i} = A_i + M_i + T_{i,n+i}$ and $B_{n+i} = A_{n+i} + a + b$, with $b = 0.5\ T_{i,n+i}$ and where a+b is also constrained to be at least 15 minutes but not more than 60 minutes.

By defining the time windows in this way, we respect the constraint on the maximum deviation from the desired service time (parameter a). Constraints on the maximum excess travel time and on maximum total travel time are relaxed, but are controlled by the sum of the parameters a+b. With this definition, the time windows are sufficiently large that the problem does not reduce to visiting a sequence of origin-destination pairs.

The solution method described in Section 2 is appropriate for any objective function which can be written as a linear function of the x_{ij} variables (distance and/or time).

2. SOLUTION METHOD.

Problem (1) subject to (2)-(10) is solved using a forward dynamic programming method. The vehicle is initially located at the departure node 0. At the first iteration, the states are made up of routes visiting a single node from among the origins. At each subsequent iteration k ($2 \leq k \leq 2n$), the states are constructed from the states of the previous iteration and are made up of routes visiting one additional node from among the origins and destinations. At the last iteration (k=2n+1), the vehicle must go to the arrival node 2n+1. We have improved the efficiency of this method, even for large-scale problems, by eliminating states (Section 3) which are incompatible with vehicle capacity, precedence and time window constraints.

2.1 Definition of states. At iteration k ($1 \leq k \leq 2n$), we define a state (S,i) if, starting at node 0, there exists a feasible route which visits all the nodes in $S \subseteq \{1,\ldots,2n\}$ and terminates at node $i \in S$; S is a non-ordered set of cardinality k.

State (S,i) is ante-feasible if there exists an order of visiting the nodes in S ending with node i which respects vehicle capacity, precedence, and time window constraints. State (S,i) is post-feasible if, beginning at node i, there exists an order of visiting the nodes in $\overline{S}$ which respects the same constraints. In the proposed algorithm, we eliminate all states which are not ante-feasible. Only some of the non-post-feasible states are eliminated: we do not eliminate states which would be more costly to identify than the savings likely to be achieved by their elimination.

2.2. Definition of labels. For state (S,i), there are several routes from node 0 to node i. These routes have the same vehicle load at node i, regardless of the order in which the nodes of S are visited: this load is denoted by y(S). However, the routes differ in terms of the arrival time at node i and the distance traveled (cost) to reach this node, depending on the order in which the nodes are visited. A given route corresponding to state (S,i) is denoted by (S_α,i). For each route, we define a (time, cost) label as $(t(S_\alpha,i), z(S_\alpha,i))$, where $t(S_\alpha,i)$ and $z(S_\alpha,i)$ represent respectively the arrival time at node i, and the distance traveled to reach this node, upon visiting the nodes of S in the order described by S_α.

At state (S,i) only some of the labels (or routes) are stored. A label is eliminated if it cannot be part of the minimum cost route from 0 to 2n+1, using the following optimality principle. Let $R = S \cup \{j\}$ and let $(t(R_\alpha,j), z(R_\alpha,j))$ be the label associated with a route from 0 to j visiting the nodes of R. If this is a minimum cost route from among all routes arriving at node j at times $t_j \leq t(R_\alpha,j)$ and if the arc (i,j) is last in this route, then the sub-route from 0 to i is a minimum cost route from among all routes arriving at node i at times $t_i \leq t(R_\alpha,j) - (M_i + T_{ij})$.

It is therefore sufficient to store the labels associated with the minimum cost routes from 0 to i visiting the set S for each arrival time i. We may eliminate for state (S,i) the labels dominated using the following partial ordering relation $\leqslant$ on the (time, cost) pairs :

$$(t,z) \leqslant (t',z') \Leftrightarrow t \leqslant t' \text{ and } z \leqslant z'. \qquad (11)$$

The label (t',z') is eliminated by (11) if there exists another label with both a lesser (or equal) time and a lesser (or equal) cost. This partial ordering relation reduces the size of the set of labels to be stored at each state (S,i). This set is given by:

$$H(S,i) = \{(t(S_\alpha,i),\ z(S_\alpha,i)),\ \alpha\varepsilon\Omega(S,i)\} \qquad (12)$$

where $\Omega(S,i)$ corresponds to the set of routes visiting the nodes in S and ending at node i, that are not eliminated.

The labels associated with a state are placed in a list with times in strictly increasing order. It can then be shown, based on the partial ordering relation, that the costs will be in strictly decreasing order. The efficient management of these lists is described in Desrosiers, Pelletier and Soumis (1983) for a dynamic programming algorithm applied to the shortest path problem with time window constraints on the nodes.

2.3. Recursion. The first iteration is carried out by visiting the origins from node 0 resulting in the set of states

$$\{(\{j\},j) \quad j\varepsilon\{1,\ldots,n\}\} \qquad (13)$$

For each state, information on the load, time and cost is easily obtained. The load is equal to the number of people to be picked up at the origin visited, and there is only one (time, cost) label associated with a state. Also

$$y(\{j\}) = C_j \qquad j\varepsilon\{1,\ldots,n\} \quad (14)$$

$$H(\{j\},j) = \{(\max\,[A_j,\ A_0 + T_{0j}],\ D_{0j}\} \qquad j\varepsilon\{1,\ldots,n\}. \quad (15)$$

At subsequent iterations, $2 \leqslant k \leqslant 2n$, new states are constructed by adding one node to the total visited at the preceding iteration. Let $(S \cup \{j\},j)$ be such a state with $j\varepsilon S$.

The load at node j is given by

$$y(S \cup \{j\}) = y(S) + C_j. \tag{16}$$

To obtain the list of labels at state $(S \cup \{j\}, j)$, we must adjust the time and cost of each label in $H(S,i)$ for all states (S,i) which are used to construct state $(S \cup \{j\}, j)$ when node j is added to the state (S,i). The new set of labels is obtained from the previous one by the function

$$f_{ij}(\{(t_i, z_j)\}) = \begin{cases} \{(\max\,[A_j, t_i + M_i + T_{ij}], z_i + D_{ij})\} \\ \quad \text{if } t_i + M_i + T_{ij} \leq B_j \\ 0 \quad \text{otherwise} \end{cases} \tag{17}$$

A new (time, cost) label can be created if starting at node i at time t_i, it is possible to arrive at node j before time B_j (the upper bound of the time window at node j). From a given state (S,i), we obtain for state $(S \cup \{j\}, j)$ the set of labels

$$F_{ij}(H(S,i)) = \bigcup_{\alpha \in \Omega(S,i)} f_{ij}(\{(t(S_\alpha, i), z(S_\alpha, i))\}) \tag{18}$$

Finally, starting with all the states (S,i) which can be used to construct the new state $(S \cup \{j\}, j)$, we obtain a new list of labels:

$$H(S \cup \{j\}, j) = \cup\, F_{ij}(H(S,i))\Vert \tag{19}$$

where the symbol $\Vert$ represents the elimination of labels according to the partial ordering relation. In the tests carried out, a reduction in the number of labels occurred at each translation by function f_{ij}.

At the final iteration, $k = 2n + 1$, there is only one state $(\{1, \ldots, 2n + 1\}, 2n + 1)$. All nodes have been visited from the departure point to the arrival point. The minimum value of the objective function is given in the label with the smallest cost, i.e. the last in the list $H(\{1,\ldots,2n+1\}, 2n+1)$ if this set is non-empty. If the set is empty, the problem is infeasible; this can be detected at any given iteration if no labels are created.

3. STATE ELIMINATION.

As the algorithm is based on dynamic programming, it is useful to limit the number of states created. For a state (S,i) to which node j is to be added, the first four elimination tests described for state $(S \cup \{j\},j)$ are independent of the terminal node i, while the next four tests depend on the labels in H(S,i). To simplify the execution of the tests, information on the states is stored in a three level structure. At the first level, we have the set S and its load y(S). At the second level, for the given set S, we have the terminal nodes $i \varepsilon S$ used to form the states (S,i). At the third level, for a state (S,i), the reduced set of labels H(S,i) is stored. (A simple example of this structure is presented in Section 4.1.)

<u>3.1. Elimination criteria based on the set S only.</u> Let S be a set at iteration k-1 to which we wish to add a node j to form state $(S \cup \{j\},j)$ at iteration k. We eliminate this state if one or more of the following criteria is not respected.

<u>#1</u> : node j must not have been previously visited : $j \varepsilon \overline{S}$.

<u>#2</u> : if j is a destination, then the origin node j-n must have been previously visited (precedence constraint): if $j \varepsilon \{n+1,\ldots,2n\}$ then $j-n \varepsilon S$.

<u>#3</u> : if j is an origin, vehicle capacity constraints must be respected: if $j \varepsilon \{1,\ldots,n\}$ then $y(S) + C_j \leqslant C$.

<u>#4</u> : time $t_j = A_j$ is the earliest time at which node j can be visited. Given this time, it must be posible to visit each subsequent unvisited node $\ell \ \varepsilon \ \overline{S \cup \{j\}}$ while respecting the time constraint:

$$A_j + M_j + T_{j\ell} \leqslant B_\ell, \quad \text{for all } \ell \ \varepsilon \ \overline{S \cup \{j\}}.$$

For all unvisited nodes, criterion #4 is not tested. The nodes are preordered in decreasing order of their upper bounds; it is reasonable to believe that this order might resemble the optimal order for visiting the nodes. The test is carried out for the most likely candidate nodes to be visited next: at iteration k+1, only the unvisited nodes ranked from k-2 to k+4 are tested. If one of these nodes cannot be visited after j, then state $(S \cup \{j\}, j)$ is rejected.

3.2. Elimination criteria based on the state (S,i). Let (S,i) be a state at iteration k-1 and t_i be the earliest arrival time at node i given by the first label in H(S,i). State $(S \cup \{j\}, j)$ will be eliminated if it is impossible to create labels. The four criteria stated formally below are based on:

- the time of the visit to node j (#5)
- the time of visit to a node subsequent to node j (#6)
- visits to two destinations whose origins are already visited, following the visit to node j (#7)
- visits to two new origins following node j (#8).

Note that criteria #1, 2, 3 and 5 determine whether a state $(S\cup\{j\}, j)$ is or is not ante-feasible, while criteria #4, 6, 7 and 8 determine whether this state is not post-feasible.

#5 : time constraint must be respected : $t_i + M_i + t_{ij} \leq B_j$.

For the three following criteria, the earliest time at which j can be visited after i is denoted by $t_j = \max [A_j, t_i + M_i + T_{ij}]$.

#6 : if we suppose that node j is visited at time t_j, it must be possible to visit each unvisited node $\ell \in \overline{S \cup \{j\}}$ while respecting the time constraints : $t_j + M_j + T_{j\ell} \leq B_\ell$, for all $\ell \in \overline{S \cup \{j\}}$. This criterion tightens criterion #4 and is tested in the same way.

#7 : if we suppose that node j is visited at time t_j, it must be possible to visit all pairs of destinations $n+\ell_1$ and $n+\ell_2$ whose origins ℓ_1 and ℓ_2 have been previously visited while respecting the time constraints. With two visiting permutations for $n+\ell_1$ and $n+\ell_2$, we must satisfy:

$$\max\,[A_{n+\ell_1},\ t_j+M_j+T_{j,n+\ell_1}] + M_{n+\ell_1} + T_{n+\ell_1,n+\ell_2} \leqslant B_{n+\ell_2}$$

or

$$\max\,[A_{n+\ell_2},\ t_j+M_j+T_{j,n+\ell_2}] + M_{n+\ell_2} + T_{n+\ell_2,n+\ell_1} \leqslant B_{n+\ell_1}$$

with $\ell_1, \ell_2 \in S \cup \{j\}$ and $n+\ell_1, n+\ell_2 \in \overline{S \cup \{j\}}$.

Criterion #7 is tested for only one pair $(n+\ell_1, n+\ell_2)$. The destinations are preordered in non-decreasing order of their upper bounds: the first two unvisited destinations whose origins have been visited are chosen.

#8 : if node j is visited at time t_j, it must be possible to visit all pairs of unvisited origins ℓ_1 and ℓ_2 while respecting the time constraints. With two visiting permutations for $n+\ell_1$ and $n+\ell_2$, we must satisfy:

$$\max\,[A_{\ell_1},\ t_j + M_j + T_{j,\ell_1}] + M_{\ell_1} + T_{\ell_1,\ell_2} \leqslant B_{\ell_2}$$

or

$$\max\,[A_{\ell_2},\ t_j + M_j + T_{j,\ell_2}] + M_{\ell_2} + T_{\ell_2,\ell_1} \leqslant B_{\ell_1}$$

with $\ell_1, \ell_2 \in \{1,\ldots,n\} \cap S \cup \{j\}$.

Criterion #8 is tested for only one pair (ℓ_1, ℓ_2). The origins are preordered in non-decreasing order of their upper bounds : the two first unvisited origins are chosen.

Note that criterion #6 (and the same applies to criterion #8) cannot be improved by also considering the destination corresponding to the unvisited origin. We would then have to satisfy:

$$\max\,[A_j,\ t_j + M_j + T_{j\ell}] + M_\ell + T_{\ell,n+\ell} \leqslant B_{n+\ell},\quad \ell,\ n+\ell \in S \cup \{j\}.$$

But the time intervals are always defined so that $B_{n+\ell} \geqslant B_\ell + (M_\ell + T_{\ell,n+\ell})$. This is a condition on the input data. Given this condition, the feasibility check for the destination follows from the check for the origin.

3.3. Elimination criterion in the case with several clients at the same location. Criterion #9, tested after criterion #5, was developed to limit the proliferation of states when i and j represent the same physical location. At a given iteration, the states $(S-\{j\},i)$ and $(S-\{i\},j)$ may exist simultaneously if the constraints are sastisfied. At the following iteration, if both states (S,i) and (S,j) are feasible, only one should be retained because the two will produce solutions of the same cost (as the distance between nodes i and j is zero).

Let i and j be the two nodes at the same location with time windows $[A_i,B_i]$ and $[A_j,B_j]$ and handling times M_i and M_j (travel times are zero). The state (S,i) or (S,j) retained is the one which has the earliest finishing time at that physical location. If t is the arrival time at this location, the state (S,j) is retained (i.e. the order $i \rightarrow j$ is imposed) if

$$\mathrm{Max}[A_j,\ \max[A_i,t]+M_i]+M_j \leqslant \max[A_i,\ \max[A_j,t]+M_j]+M_i. \quad (20)$$

The left-hand side of (20) corresponds to the order $i \rightarrow j$ and conversely for the right-hand side. It can be verified that this inequality is valid if $A_i \leqslant A_j$. This condition depends partly on the time t since both states (S,i) and (S,j) have to be feasible in order to make a choice. If $A_i = A_j$, $i \rightarrow j$ or $j \rightarrow i$ may be chosen arbitrarily. If the vehicle capacity is constraining (which was not the case in our test problems), relation (20) is sufficient to impose the order $i \rightarrow j$ in the case where the nodes are two origins or two destinations. When one is an origin and the other is a destination, customers must be dropped off before new ones can be picked up, i.e. node i must be a destination and node j be an origin.

The reader should note that this criterion does not eliminate states generated from $(S-\{j\},i)$ or $(S-\{i\},j)$ and other nodes which are not at the same physical location as i and j. So, if this is optimal, the vehicle may only handle part of the service at a given physical location at the first visit, and it may then return to serve the rest.

This criterion reduces the number of states generated in the following way. (An example is presented in Figure 1 with nodes i, j and k and $A_i \leq A_j \leq A_k$.)

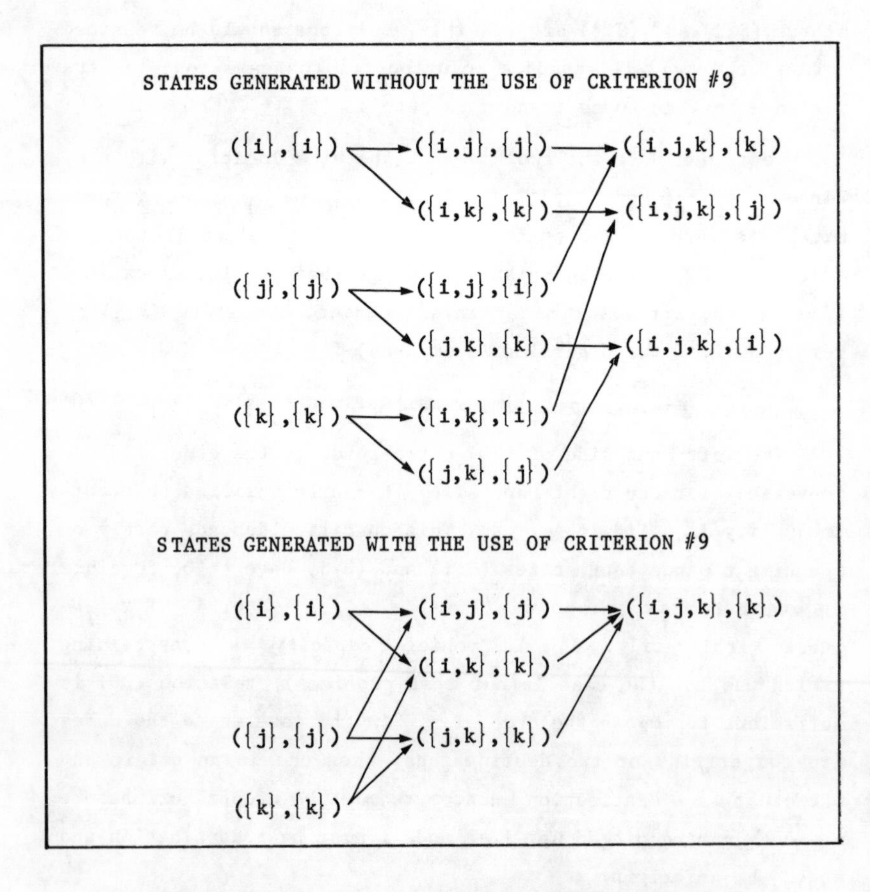

Figure 1 : Effect of Elimination Criterion #9 on the Number of States Generated.

Suppose v nodes are located at the same place. Each time it is possible to visit these v nodes in the algorithm in any order, the number of states generated during the v consecutive iterations is given by:

$$C_1^v + C_2^v + \dots + C_v^v = 2^v - 1 \tag{21}$$

using criterion #9. If criterion #9 is not used, this number becomes

$$vC_1^{v-1} + vC_1^{v-1} + \dots + vC_{v-1}^{v-1} = v2^{v-1} \tag{22}$$

Criterion #9 thus produces a reduction by a factor of approximately 2/v in the number of states generated.

Finally, criterion #9 may be very useful when several nodes are very close to one another without being at exactly the same place. A simple way to limit the proliferation of states is to group the nodes at the same place and apply this criterion. The loss in optimality will not be great, since the distances involved are very small, but computation times may be significantly reduced.

4. NUMERICAL RESULTS.

In this section, we illustrate how the algorithm works, with a simple example, give certain details concerning its implementation, and finally present the results of the tests carried out.

4.1. Example. Figure 2 shows an example with only two requests, where transportation is required from node 1 to node 3, and from node 2 to node 4. The time interval during which each node is to be visited is indicated on the diagram in Figure 2. To simplify the presentation, we assume that the vehicle capacity will not be violated, and that the number associated with each arc represents simultaneously the cost and the travel plus handling times. At the bottom of Figure 2, the three level information structure is illustrated: the sets and terminal nodes from the states, while the (time, cost) labels are at the third level.

At the first iteration, the origins are visited and states

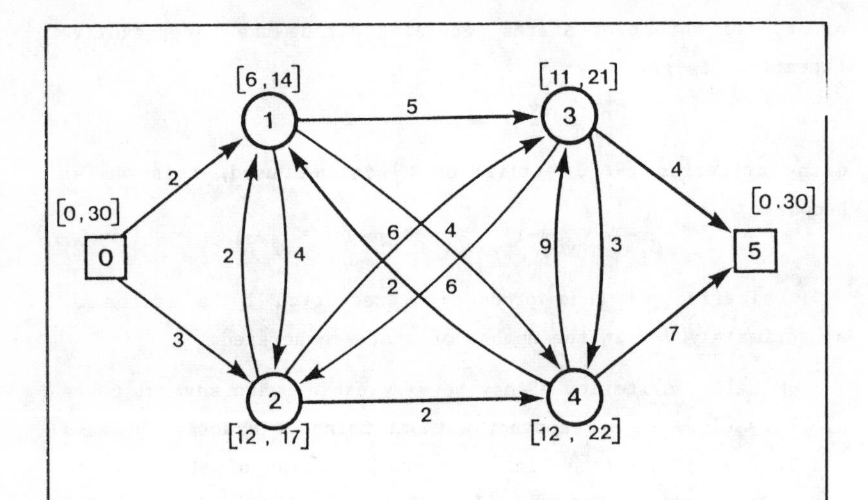

k	STATES			LABEL'S NUMBERS	PRECEDING LABEL NUMBER
	SETS	TERMINAL NODES	LABELS		
1	{1} {2}	1 2	(6,2) (12,3)	1 2	∅ ∅
2	{1,2} {1,3}	1 2 3	(14,5) (12,6) (11,7)	3 4 5	2 1 1
3	{1,2,3}	2 3	(17,13) (18,12) (19,10)	6 7 8	5 4 3
4	{1,2,3,4}	4	(19,15) (22,13)	9 10	6 8
5	{1,2,3,4,5}	5	(26,22) (29,20)	11 12	9 10

Optimal route : 5←4←3←1←2←0 : Cost = 20; Time = 29

Alternative route : 5←4←2←3←1←0 : Cost = 22; Time = 26

Figure 2 : Numerical Example.

({1},1) and ({2},2) are created with one label each. At iteration k=2, the feasible states are constructed either from one origin and its destination (i.e. state ({1,3},3) with a single label), or from two origins (i.e. ({1,2},1) and ({1,2},2) with one label each. The infeasible states are eliminated as follows: by criterion #1, states ({1},1) and ({2},2); by criterion #2, states ({1,4},4) and ({2,3},3); and by criterion #6, state ({2,4},4) as it is impossible to visit node 1 afterwards while satisfying the time window constraint. At iteration k=3, state ({1,2,3},2) is created with the label (17, 13) as well as state ({1,2,3},3) with two labels (18,12) and (19, 10) placed in decreasing time order. State ({1,2,4},4) is eliminated by criterion #6 as it is impossible to visit node 3 afterwards while satisfying the time window constraint. At iteration k=4, only one state can be constructed from the states of iteration 3, and adjusting the previous labels with numbers 6, 7 and 8, we obtain only two labels as follows :

$$(17,13) \xrightarrow{+2} (19,15)$$

$$(18,12) \xrightarrow{+3} (21,15) \text{ eliminated by label } (19,15)$$

$$(19,10) \xrightarrow{+3} (22,13)$$

Finally, we obtain two routes at iteration k=5 : the optimal route has a cost of 20 and finishes at time 29, while the quickest route takes a time of 26, at a cost of 22. The routes are identified by backtracking the label numbers.

Note that each (time, cost) label corresponds to an ordered visit to a set of nodes. The three level structure proposed uses less space and simplifies the use of the state elimination criteria.

4.2. Implementation. We now describe certain details of the implementation of the numerical tests. The initialization of the algorithm is carried out directly at iteration k=2 by considering all distinct origin pairs $i \neq j$, $1 \leq i, j \leq n$ satisfying criteria

#3 to #9, and all origin-destination pairs i and n+i satisfying criteria #6 to #8. The cost between two nodes is given in minutes. In fact, the only values stored in a non-symmetric matrix are the $T_{ij} + M_i$ (travel time between i and j plus handling time at node i). These same values are used for the costs $D_{ij} = T_{ij} + M_i$, as the vehicle visits all the nodes and the sum of handling times is a constant.

The data structure with three levels is retained in a packed form. One word per set contains the first level information: the feasible set, and a pointer to the first terminal node. One word per state contains the second level information: terminal node, load, pointer to the following terminal node for this same set and a pointer to the first label. One word per label contains third level information: time, cost, pointer to the state (S,i), and two pointers to the preceding and following labels in the list H(S,i).

To limit memory space requirements, at iteration k, we store at the first and second levels only the feasible sets and terminal nodes of the current iteration k and the preceding iteration k-1. Furthermore, at the end of iteration k, the labels of iteration k-1 which were not used to generate new labels are eliminated from the structure.

<u>4.3. Tests</u>. The 98 problems investigated had from 5 to 40 requests and were derived from the cities of Montréal, Toronto and Sherbrooke in Canada. The problems with over 25 requests were constructed by combining smaller problems based on morning, afternoon, and evening routes. The main characteristics of these problems are presented in <u>Table 1</u>. The time windows associated with the origins and destinations are of the order of 40-50 minutes, with a minimum of 30 minutes and a maximum of 82 minutes. The "route diameter" is a measure of the size of the area covered by a route. It is expressed in terms of the travel time along a straight line joining the two most distant points

Table 1: Problem characteristics.

CITY	NUMBER OF PROBLEMS	TIME WINDOWS[1]			ROUTE DIAMETER[2]		MEAN ARC SIZE[2]
		MIN	MEAN	MAX	MEAN	MAX	
MONTREAL	65	30	38	51	44	70	12
SHERBROOKE	9	30	38	59	24	33	9
TORONTO	24	40	53	82	43	63	13

[1] In seconds.
[2] Travel time in minutes.

visited by the route. This diameter is on average 43-44 minutes for the large cities (Montreal and Toronto), and around 24 minutes for Sherbrooke, a small town. The last column gives the average travel time between two consecutive points along the optimal routes. Note that the time windows are large compared with the mean arc size of around 10 minutes. The problems tested therefore may have a very large number of feasible routes; each point has on average 6 possible successors respecting the time windows. In addition, vehicle capacity was always sufficient and all problems were feasible.

Figure 3 shows the execution time in seconds as a function of problem size. Test problems were solved on the University of Montreal Cyber 173 computer using the FORTRAN 5 (opt=2) compiler. Almost all (76 out of 82) of the original problems with $n \leq 25$ were solved in less than two seconds. Over the whole set of problems, the relationship between execution time and problem size is almost linear. This relationship can be expected as there is no interaction between two requests separated by several hours. Tests on additional larger scale randomly generated feasible problems produced similar results.

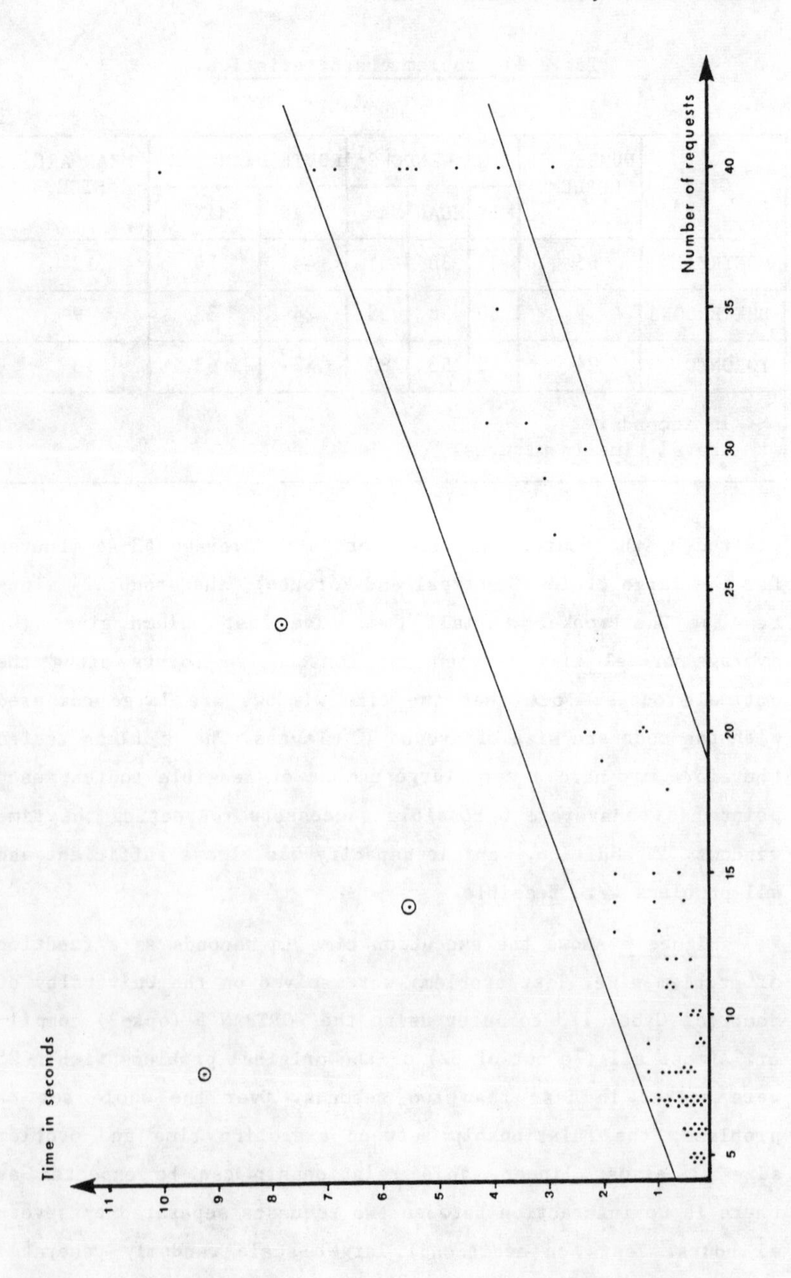

Figure 3: Execution Time as a Function of the Number of Requests

For the more detailed study of the numerical results, we excluded three problems which are easily identifiable in Figure 3 as having high execution times (with n=8, 14 and 24). These problems have origins and/or destinations which are close together for several requests, with very similar time windows. For example, for the excluded problem with n=8, each node has 11 possible successors on average while respecting the time constraints. This results in a proliferation of states as occurs in the treatment of the traveling salesman problem by dynamic programming.

Table 2 presents various averages by class and size of problem. We noted above the linear relationship between execution time and problem size. The graphical representation of the number of states generated also shows the same relationship. The number of terminal nodes per feasible set is relatively small, less than two on average. The label elimination process using the partial ordering relation reduces the number of labels to be stored for each state to two on average. Even if the elimination is not very strong, it avoids the generation of all successors of these labels and it produces a great effect. The proportional relationship between the number of states generated and the problem size n indicates that the average number of states per iteration is almost constant. Dividing by 2n iterations, we obtain on average five to seven states per iteration. The memory used is thus small. Without including the matrix structure of the values ($T_{ij} + M_i$), 97 of the 98 problems handled were solved with less than 1500 words, the exception being the difficult problem of size n=8 with 2661 words.

Table 3 shows elimination statistics on the criteria described in Section 3. Criteria #1 and #2 (new node and precedence), which are not really tests but just part of the definition of the successor states, eliminate over 60% of the states in general. The effect of other elimination criteria has been

Table 2: Summary of Results of 95 Problems.

SIZE n	NUMBER OF PROBLEMS	STATES[1]	TERMINAL NODES[2]	LABELS[3]	EXECUTION TIME[4]
5-7	36	57	1.54	1.14	0.31
8-10	18	86	1.59	1.24	0.47
11-19	19	194	1.70	1.37	1.23
20-30	10	287	1.75	1.25	2.57
31-40	12	431	1.72	1.28	5.27

[1] Average total number of remaining states per problem.
[2] Average number of terminal nodes per feasible set.
[3] Average number of labels per state.
[4] Average execution time in seconds (CYBER 173).

presented in the order of their application (4, 5, 9, 6, 7, 8). Each column indicates the percentage of states tested with the criterion which have been eliminated (example: for size 31-40, the average total number of states was 23869 + 431 = 24300, criteria #1 and #2 eliminated 17010 states; the value 70% is calculated by the ratio 17010/2400, the 7290 remaining states were tested by criterion #4 and 6342 were eliminated giving 87% success based on the ratio 6342/7290, ...).

Table 3: Elimination statistics by criterion.

SIZE n	STATES[1] ELIMINATED PER PROBLEM	#1 & #2 (2)	#4 (2)	#5 (2)	#9 (2)	#6 (2)	#7 (2)	#8 (2)
5-7	470	65	22	2	9	43	5	11
8-10	910	65	45	5	7	47	8	5
11-19	3058	69	63	6	7	41	7	5
20-30	6936	69	70	13	13	49	8	7
31-40	23869	70	87	18	18	45	6	3

[1] Average number of states eliminated by problem.
[2] Percentage of states tested with this criterion which have been eliminated.

Criterion #4 (visit to one additional node if node j is visited at time A_j) which test "post-feasibility" is the most important and the performance increases with problem size (from 22% to 87%).

The performance of criterion #5 (time window constraint) also increases with problem size (from 2% to 18%). The percentage of states eliminated by criterion #9 (same location) increases with problem size (from 9% to 18%) and is dependent on the percentage of multiple stops at the same physical location in the test problems. Criterion #6 (visit to one additional node if node j is visited at time t_j) is very important for every problem size (from 40% to 50%). The performance of criteria #7 and #8 (visit to two additional nodes) is less important while criterion #3 (capacity) is inactive.

5. CONCLUSION.

Our algorithm easily solves problems with 40 requests. The corresponding traveling salesman problem has 80 nodes and could certainly not be solved by dynamic programming (see Bellmore and Nemhauser (1983)). The effectiveness of our dynamic programming algorithm is largely due to the use of efficient elimination criteria for states which are infeasible because of the additional constraints on the route. Problems with over 40 requests were not tested because, in the context of transportation for the handicapped, a vehicle rarely handles more than around twenty requests in a day. As the number of states and the computation times are small and as these increase linearly with the number of requests, and as the memory required is very small and does not increase with the number of requests, much larger problems could in fact be solved.

This algorithm is currently being used for the optimization of routes produced by a heuristic for the problem with several vehicles. The distance traveled was reduced by 4% to 5% without diminishing the quality of the service. The algorithm could also be used to verify the feasibility of serving a set of requests with one vehicle in a "cluster first-route second" approach.

ACKNOWLEDGMENTS

The authors wish to thank the Federal Government of Canada (NSERC and PUT programs), the Quebec Government (FCAR program), and L'École des Hautes Études Commerciales de Montréal, for their financial support.

REFERENCES

Armstrong, G.R. and Garfinkel, R.S. (1982). Dynamic programming solution of the single and multiple pick-up and delivery problem with application to dial-a-ride. Working Paper #162, College of Business Administration, University of Tennessee, Knoxville, TN.

Bélisle, J.-P., Soumis, F., Roy, S., Desrosiers, J., Dumas, Y. and Rousseau, J.-M. (1984). The impact on vehicle routing of various operational rules of a transportation System for handicapped persons. Publication #363*, Centre de recherche sur les transports, Université de Montréal, Montréal, Canada H3J 3J7. (Presented at the Third International Conference on Mobility and Transport of Elderly and Handicapped Persons, Orlando, Florida).

Bellemore, M. and Nemhauser, G.L. (1968). The travelling salesman problem - a survey. Operations Research 16, 538-558.

Bodin, L., Golden, B., Assad, A. and Ball, M. (1983). Routing and scheduling of vehicles and crews - The state of the art, Computers and Operations Research 10, 63-211.

Bodin, L. and Sexton, T. (1986). The multi-vehicle subscriber dial-a-ride problem. The Delivery of Urban Services, TIMS Studies in the Management Science 22, forthcoming.

Desrosiers, J., Dumas, Y. and Soumis, F. (1984). The multiple vehicle many to many routing problem with time windows. Publication #362, Centre de recherche sur les transports, Université de Montréal, Montréal, Canada H3C 3J7.

Desrosiers, J.. Pelletier, P. and Soumis, F. (1983). Plus court chemin avec contraintes d'horaires, R.A.I.R.O. 17, 4, 357-377.

Jaw, J., Odoni, A., Psaraftis, H. and Wilson, N. (1984). A heuristic algorithm for the multi-vehicle advanced request dial-a-ride problem with time windows. Working Paper MIT-UMTA-83-L, MIT, Cambridge, Ma 02139.

Psaraftis, H. (1980), A dynamic programming solution to the single vehicle many to many immediate request dial-a-ride problem. Transportation Science 14 (2), 130-154.

Psaraftis, H. (1982). An exact algorithm for the single vehicle many to many dial-a-ride problem with time windows, Technical Note OE-UMTA-82-2, M.I.T., Cambridge, MA 02139.

Psaraftis, H. (1983). Analysis of an $O(N^2)$ heuristic for the single vehicle many-to-many euclidian dial-a-ride problem, Transportation Research B 17B (2), 133-145.

Roy, S., Rousseau, J.-M., Lapalme, G. and Ferland, J.A. (1984a). Routing and scheduling for the transportation of disabled persons - The algorithm. TP 5596E. Transport Development Centre, Transport Canada Montréal, Canada H2Z 1X4.

Roy, S., Rousseau, J.-M., Lapalme, G. and Ferland, J.A. (1984b), Routing and scheduling for the transportation for disabled persons - The tests. TP 5598E. Transport Development Centre, Transport Canada, Montréal, Canada H2Z 1X4.

Sexton, T. and Bodin, L. (1985a), Optimizing single vehicle many-to many operations with desired delivery times: I. Scheduling. Transportation Science 10 (4), 378-410.

Sexton, T. and Bodin, L. (1985b), Optimizing single vehicle many-to many operations with desired delivery times: II. Routing. Transportation Science 10 (4), 411-435.

Received 11/84; Revised 6/28/86.

AMERICAN JOURNAL OF MATHEMATICAL AND MANAGEMENT SCIENCES

SCHEDULING LARGE-SCALE ADVANCE-REQUEST DIAL-A-RIDE SYSTEMS

Harilaos N. Psaraftis
Massachusetts Institute of Technology
Cambridge, Massachusetts 02139

SYNOPTIC ABSTRACT

This paper examines the scheduling of large-scale advance-request dial-a-ride systems, describes two algorithms that have been developed recently to solve problems in this area, and provides analysis and insights into how these algorithms are expected to perform under various operational scenarios and in comparison with one another. The algorithms examined are the GCR (Grouping/Clustering/Routing) and ADARTW (Advanced Dial-A-Ride with Time Windows) procedures. The paper gives an overview of both algorithms, emphasizes the differences in their operational scenarios, describes computational experience with both procedures and includes worst-case considerations for ADARTW. Extensions and directions for further research are also discussed.

Key Words and Phrases: Scheduling with Time Windows; Vehicle Routing; Dial-A-Ride Systems.

1986, VOL. 6, NOS. 3 & 4, 327-367
0196-6324/86/030327-41 $11.20

1. INTRODUCTION

This paper examines the scheduling of large-scale advance-request dial-a-ride systems, describes two algorithms that have been developed recently in the above context, and provides analysis and insights into how these algorithms are expected to perform in various operational scenarios and in comparison with one another.

The problem we are concerned with is that of routing and scheduling a fleet of vehicles to serve customers who have to be picked up from specified origins and delivered to specified destinations. Our operating scenario assumes that all of the requests for service are received well in advance of the actual time of vehicle dispatching (say, the day before at the latest). It also assumes that each customer has specified a desired pickup time or a desired delivery time (but not both) and that the vehicle operating agency has adopted some guidelines so as to ensure adequate quality of service to all customers. Such guidelines, also known as customer service guarantees (or service quality constraints), are adopted so as to avoid "intolerable" deviations from each customer's desired (pickup or delivery) times and/or excessively circuitous routes that would force a customer to spend "too long" a time on board the vehicle (as compared to that customer's direct ride time). As we shall see, such customer service guarantees are typically translated into "time-windows" for each customer's pick up and/or delivery times, in a fashion that will be explicitly defined later. At the same time, the scheduler would also like to achieve a good utilization (or productivity) for the fleet. Given that such a goal is generally in conflict with the objective of an acceptable quality of service, the problem is to find the best way to utilize available resources so that both goals are "satisficed".

The generic problem described above is known by several

names in the Transportation/Operations Research literature, such as "multi-vehicle many-to-many advance-request dial-a-ride problem", "subscriber dial-a-ride problem", "dial-a-ride problem with desired pickup or delivery times", "dial-a-ride problem with time windows", and so on. This problem is to be contrasted with the equivalent "immediate-request" problem, in which customer requests are received and processed at the same time as the actual time of vehicle dispatching (that is, in real time). Most real-world dial-a-ride agencies operate a "mixed" service, where both types of requests (in varying proportions) can be handled simultaneously. In this paper we shall be concerned with a pure advance reservation configuration only.

There has been an abundance of algorithmic development efforts in this area in the recent past. Wilson and Weissberg (1976) and Wilson and Colvin (1977) developed heuristic algorithms for the dial-a-ride system of Rochester, New York. Sexton (1979) and Bodin and Sexton (1982) developed single vehicle and multi-vehicle approximate algorithms which they applied to the subscriber dial-a-ride system of Baltimore, Maryland. Psaraftis (1980) described an exact approach for the single-vehicle immediate-request problem, and modified this approach for the equivalent problem with time windows (Psaraftis (1983a)). He also developed several polynomial-time heuristics for the single-vehicle immediate-request problem (Psaraftis (1983b, 1983c)). (A polynomial-time algorithm is one whose running time is bounded by a polynomial function of the size of the problem). Hung, Chapman, Hall and Neigut (1982) developed a procedure (also known as the "Neigut/NBS" algorithm) for the multi-vehicle time window problem, and Roy, Chapleau, Ferland, Lapalme and Rousseau (1983) described an algorithm for the same version of the problem. Finally, the current author and his colleagues developed two different multi-vehicle advance-request dial-a-ride algorithms during the past 2-3 years: in Jaw, Odoni,

Psaraftis and Wilson (1982), the "Grouping/Clustering/Routing" (or GCR) algorithm was described, while in Jaw, Odoni, Psaraftis and Wilson (1984) the "Advanced Dial-A-Ride algorithm with Time Windows" (or ADARTW) was introduced.

The focus of this paper is on the two most recent large-scale dial-a-ride algorithms, GCR and ADARTW. Specifically, Section 2 gives an overview of both algorithms and illustrates the main similarities and differences with regard to the operating scenario, the treatment of time constraints, the solution approach, and other features. Section 3 describes some recent, encouraging computational experience with the two procedures, including some comparisons of the two using the same data set. Section 4 concentrates on ADARTW, by presenting some worst-case considerations and related scenarios. The analysis identifies the features in the structure of ADARTW that are likely to cause undesirable schedules in some (rare) cases. Finally, Section 5 contains some brief concluding remarks on directions for further research, including an extension to the "mixed" demand case.

2. OVERVIEW OF THE GCR AND ADARTW APPROACHES

Although the real-world problems for which GCR and ADARTW have been designed are essentially the same, there are some differences in the way "reality" is modeled in each of the two approaches. In general, conceptual terms, the GCR approach assumes that a customer would be satisfied if he is picked up or delivered "reasonably close" to his specified pickup or delivery time, whereas the ADARTW approach provides a much stricter definition of what constitutes an acceptable level of service. Moreover, the ADARTW approach assumes that the dispatcher considers vehicle fleet size as a decision variable, whereas in GCR fleet size is treated as a given parameter. The rest of this section discusses these and other more subtle differences in detail. For reference purposes, a list of symbols and abbreviations used throughout the text appears in Table 1, and a summary of assumptions and features of GCR and ADARTW is listed in Table 2.

2.1 Type-P and Type-D Customers. Both approaches assume that each customer requesting service has specified either a desired pickup time (DPT) or a desired delivery time (DDT), but not both. Such an assumption is reasonable in view of the fact that customers using the system are likely to be time-constrained only on one end of their trip (usually the delivery end during the morning and the pickup end during the afternoon). ADARTW actually goes further by assuming that each DPT is the earliest time the corresponding customer can be picked up (EPT) and that each DDT is the latest time the corresponding customer can be delivered (LDT). Both approaches allow a mix of both categories of customers (from now on referred to as type-P and type-D respectively) in the same problem. Also, both approaches can be modified in a straightforward way to allow a customer to specify both a DPT and a DDT. In fact (also see Section 2.2) both approaches derive, each in a different way, "equivalent" delivery (pickup) times - or time windows - for each type-P (type-D) customer. These derivations would be bypassed if a customer happens to specify both a DPT and a DDT, and these values would be used directly by the algorithms. Of course, in this case separate checks should be applied to ascertain whether the computer-specified DPT and DDT values constitute a feasible requirement (also see Section 2.2).

2.2 Time Constraints and Service Guarantees. Perhaps the most important difference between the operating scenarios assumed by the two approaches is the treatment of time constraints and the corresponding customer service guarantees. For each type-P(D)

TABLE 1: List of Symbols and Abbreviations

a,b	Constants Specifying MRT as Function of DRT
ADARTW	Advanced Dial-A-Ride (Algorithm) with Time Windows
ADT	Actual Delivery Time
APT	Actual Pickup Time
ART	Actual Ride Time
AVF:	Active Vehicle Fleet
BVF	Backup Vehicle Fleet
$c_1,\ldots,c_8$	Objective Function Coefficients in ADARTW
CF	Conversion Factor in GCR
$COST_{ik}$	Minimum Marginal Insertion Cost of Customer i into Work Schedule of Vehicle j
d_{ij}	Direct Distance from i to j
D(i,j)	Direct Trip Time from i to j
DDT	Desired Delivery Time
DPT:	Desired Pickup Time
DRT	Desired Ride Time
DT	Time Group Length in CCR
DU_i, DU_i	Disutility to Customer i
EDT	Earliest Delivery Time
EPT	Earliest Pickup Time
GCR	Grouping/Clustering/Routing (Algorithm)
i,j,k	Indices for Points, Customers, Clusters
K	Pool Size
ln	Natural Logarithm
LDT	Latest Delivery Time
LPT	Latest Pickup Time
m	Index for Vehicles
MC_{ij}	Marginal Insertion Cost of Customer i into Work Schedule of Vehicle j
$MINCOST_i$	Minimum of $COST_{ij}$ (w.r. to j)
MRT	Maximum Ride Time
n	Number of Customers, Size of UCL
S	Work Schedule of Vehicle
SW_i	System Workload Index for Customer i
T	Constant Used to Calculate SW_i
type-D	Customer Having Specified a DPT
type-P	Customer Having Specified a DDT
UCL	Unprocessed Customer List
VC_i	Cost of Vehicle Resources Due to Inserting Customer i
x_i	Pickup or Delivery Time Deviation of Customer i
y_i	Excess Ride Time of Customer i
W	Time Window Width
w_i	Additional Slack Vehicle Time Due to Inserting Customer i
z_i	Additional Active Vehicle Time Due to Inserting Customer i
Z_{ADARTW}	Objective Function Value Produced by ADARTW
Z_H	Objective Function Value of Heuristic H
Z_{OPT}	Optimal Value of Problem
Z_R	Reference (Maximum) Objective Value
$\Omega(n)$	Growing at Lease as a Linnear Function of n

TABLE 2: Summary of assumptions and features of GCR and ADARTW algorithms and their operating scenarios. Asterisks (*) denote nonbinding assumptions that can be relaxed without major modifications in the logic of the algorithm.

ASSUMPTION/FEATURE	GCR ALGORITHM	ADARTW ALGORITHM
Demand Environment	Pure Advance Request	Pure Advance Request
Customer Desired Service Times	Each customer specifies either a desired pickup time (DPT) or a desired delivery time (DDT) but not both	Same as GCR. In addition, each customer's DPT (DDT) is equal to that customer's earliest pickup (latest delivery) time EPT (LDT).
Customer Type Mix	Both DPT and DDT-specified customers can exist in same problem	Same as GCR
Customer Service Guarantees	<u>Soft</u>: Attempts (with no guarantees) to service each customer within a time interval of specified duration. Non-empty vehicles cannot idle.	<u>Hard</u>: Accepts upper bounds on each customer's deviation from desired (pickup or delivery) time <u>and</u> on each customer's ride time. Non-empty vehicles cannot idle.
Customer Dwell Times	Zero	Zero (*)
Distance Matrix	Symmetric	General
Vehicle Fleet Size	A User-Input. Can vary in time.	A Decision Variable. Can vary in time.
Vehicle Types	Identical	Identical (*)
Vehicle Capacity Constraint	No	Yes
Options if Service Guarantees Cannot Be Met With Available Resources	Problem never infeasible. If service deteriorates, scheduler may either deny service or iterate with more vehicles on a trial/error basis.	Can either deny service to some customers or add more vehicles at a cost.
Objective Function	Not explicitly defined. Attempts to equalize vehicle workload, maximize total vehicle productivity and minimize route circuity.	Minimizes a prescribed function of customer disutility and vehicle resource utilization. User-calibrated via eight coefficients
Solution Method	Grouping-Clustering-Routing: Customer decomposition in time; seed customer selection; simultaneous assignment of other customers; look-ahead capability; use of a routing algorithm as subroutine.	Sequential Insertion: Includes fast screening test for feasible insertions and optimization of customer assignments.
Worst-Case Performance	Not Applicable	Worst-Case Relative Error: + oo Worst-Case Absolute Error: Unknown
Average Case Performance	Not Applicable	Unknown
Computational Complexity	$0\ (mn + n^2)$ for n customers and m vechicles on the average	$0\ (n^4)$ in worst case
Maximum Problem Size Solved	2,617 customers	2,617 customers
CPU Time for Max. Problem VAX 11/750	6-7 mins.	12 mins.
Computer language used	FORTRAN	PL1
Extension to "Mixed" Demand Case	Difficult	Easier

customer, the GCR algorithm simply derives an equivalent desired delivery (pickup) time DDT(DPT), as a function of that customer's desired pickup (delivery) time DPT(DDT) and his/her direct ride time DRT as follows (see also Figure 1):

$$\text{For type-P customers, } DDT = DPT + CF * DRT \quad (1)$$

$$\text{For type-D customers, } DPT = DDT - CF * DRT \quad (2)$$

where CF (for "conversion factor") is a calibration parameter ($\geq$1) to account for possible circuity in that customer's ride. (A high value of CF means that the difference DDT-DPT is greater than DRT, hence the customer may arrive to his/her destination via a circuitous route) GCR then divides the entire time horizon of the problem into adjacent "time groups" (see Figure 1). Each time group has a duration of DT, with DT being another calibration parameter. Sensitivity on the selected values of CF and DT is reported in Section 3.

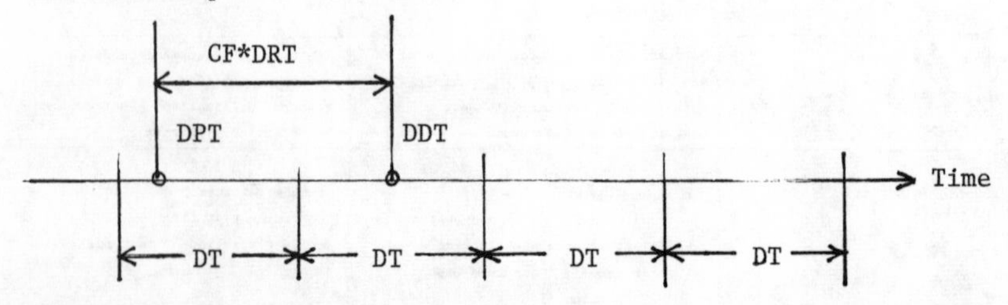

Figure 1: Time groups in GCR algorithm

In GCR, customer service guarantees (and corresponding time constraints) are "soft" in the sense that the algorithm will attempt to service (pickup or deliver or both) each customer only within the time group that encompasses his DPT or DDT, without directly considering that customer's individual desired time. For instance, if time groups are set up every 30 minutes starting at 8:00 a.m., an attempt will be made to pick up a customer who has specified a DPT of 10:37 a.m. between 10:30 a.m. and 11:00 a.m., with the projected actual pickup time being quoted to him well in advance of actual vehicle dispatching.

The operating scenario of ADARTW is rather different (see Figure 2).

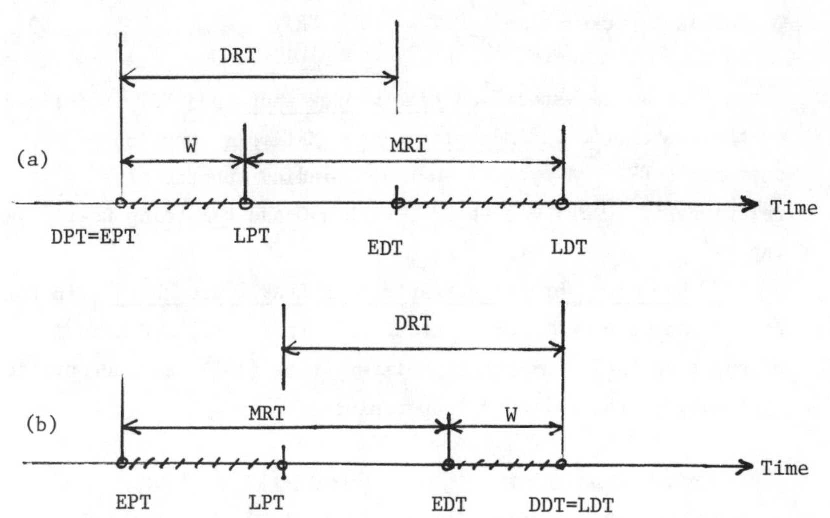

Figure 2: Time windows in ADARTW algorithm, (a) for type-P customers, (b) for type-D customers.

For each type-P(D) customer ADARTW derives three types of times, which, together with that customer's EPT (LDT), constitute a pair of time windows (one for pickup and one for delivery) for that customer. These times are defined as follows:

For type - P customers (EPT specified, see Figure 2a):

Latest pickup time $LPT = EPT + W$ (3)

Earliest delivery time $EDT = EPT + DRT$ (4)

Latest delivery time $LDT = LPT + MRT$. (5)

For type - D customers (LDT specified, see Figure 2b):

Earliest delivery time $EDT = LDT - W$ (6)

Latest pickup time $LDT = LDT - DRT$ (7)

Earliest pickup time $EPT = EDT - MRT$, (8)

where W is a user-specified *time window width* and MRT is defined as the customer's *maximum ride time*. MRT is a function of a customer's DRT. A typical, but not binding functional relationship is $MRT = a + b*DRT$, with $a \geq 0$ and $b \geq 1$ being user inputs.

Customer service guarantees in ADARTW are "hard", in the sense that each customer's actual ride time (ART) and actual pickup time (APT) (or actual delivery time (ADT), as appropriate) *must* satisfy the following constraints:

for all customers: $DRT \leq ART \leq MRT$ (9)

for type-P customers only: $EPT \leq APT \leq LPT$ (10)

for type-D customers only: $EDT \leq ADT \leq LDT$ (11)

where LPT in (10) and EDT in (11) are defined by (3) and (6) respectively.

It is easy to check that *if* (9) and (10) ((9) and (11)) are satisfied for a type-P(D) customer, then that customer's ADT (APT) must lie between EDT(EPT) and LDT (LPT) as defined by (4) and (5) ((8) and (7)) respectively (the opposite is not necessarily true).

If we allow a customer to specify both a DPT and a DDT, then both these values would be used as such in the GCR algorithm, that is, the conversion implied by (1) or (2) would not be used. However, in this case we should also require that the ratio

(DDT-DPT)/DRT be greater than 1.0 (otherwise the request would be infeasible). In ADARTW, if a customer has specified both a DPT and a DDT, we would use EPT=DPT, LDT=DDT, and derive LPT and EDT from (3) and (6) respectively. In this case, the conversions implied by (4), (5), (7) and (8) would not be used. Also, in this case the maximum ride time constraints would be automatically satisfied whenever DPT-DPT a+b*DRT.

The issue of which of the two approaches ("soft" or "hard") is more appropriate, has been and is being widely debated, both among algorithm developers and among operators and policy makers. We feel that each approach has merit, as well as drawbacks; these are discussed further in Section 5.

2.3 Rejecting Customers versus Adding More Vehicles. Since the "hard" approach in ADARTW raises the prospect of infeasibility (which does not exist in GCR due to the absence of such constraints), there are two basic user options by which this issue can be handled: Either by denying service to those customers whose service guarantees cannot be met, or by adding more vehicles (drawn from a "backup vehicle fleet"), at a cost. This cost is explicitly considered in ADARTW's objective function by a term that quantifies the "cost of vehicle resources" (see Section 2.6). Such decisions can also be made in GCR, but only at the discretion of the dispatcher. That is, whenever he feels that some customers suffer from poor quality of service, or whenever the capacity of some vehicle is exceeded (GCR does not consider vehicle capacity constraints explicitly), the dispatcher has the option to deny service to some customers, or rerun GCR with more vehicles until service quality becomes acceptable and capacity constraints are satisfied..

2.4 Optimizing Vehicle Fleet Size. The treatment of vehicle fleet size as a function of time is another major difference between the two approaches. In GCR, the number of vehicles to be operated throughout the day is supplied by the user, possibly as a function of customer demand (this number can vary over time). In ADARTW, this number is a decision variable, determined so as to optimize a prescribed objective function which includes vehicle resource costs (see also equation (16) below).

2.5 Objective Function and Solution Method/GCR. The ultimate objective in GCR is to maximize total vehicle productivity (in terms of passengers serviced per vehicle hour). However, the design of the algorithm is such that the above objective is never explicitly considered by the procedure. Instead, GCR produces a set of routes and schedules by attempting to optimize a set of surrogate measures, defined in such a way that the resulting solution is likely to achieve good productivity and a good quality of service. The general solution method of GCR, along with the performance measures that the algorithm attempts to optimize at each step, is summarized as follows (for complete details see Jaw, Odoni, Psaraftis and Wilson (1982)).

STEP 1 (Grouping): Based on each customer's DPT and DDT (one of them being input and the other calculated by (1) or (2)), and on user inputs DT and CF, divide the time horizon into equal and consecutive time groups, and then assign customers to those groups, according to the interval into which their DPT's or DDT's fall. This is just a decomposition of the problem by time, and no optimization is involved.

STEP 2 (Clustering): For each time group defined in Step 1, further subdivide all customers of that time group into "clusters", and then assign a vehicle to service (pickup, deliver, or both) all customers in each cluster. The mechanism by which this is done is fairly intricate (see Jaw, Odoni,

Psaraftis and Wilson (1982)) but can be described roughly as follows:

For each time group k do the following:

Step 2.1: Let m be the number of vehicles available (but thus far unassigned) in k. Identify m unassigned customers in k and proclaim each of them a "seed" customer for a new cluster in k. Do this by attempting to optimize a measure of "seed dispersion", that is, declare a customer a "seed" if he maximizes the "distance" between him and the closest seed already chosen. "Distance" between two seeds is defined as a function of the distances between origins and destinations of the corresponding customers, depending on the scenario.

Step 2.2: Assign the m available vehicles to the m seed customers obtained in Step 2.1. Do this by attempting to minimize the total cost of the assignment. The cost associated with each vehicle-seed pair is defined as the distance between the vehicle and the origin of that seed.

Step 2.3: Add all other customers in group k, one by one, to either the new clusters formed in Step 2.1, or, to already existing clusters carried from group k-1. Each cluster is allowed to grow in such a way so that "vehicle workload" is spread out evenly among all vehicles. The vehicle workload computation includes some "look ahead" features to take into account the locations of the origins and destinations of customers belonging to time group k+1.

STEP 3 (Routing): For each time group k and for each cluster identified in Step 2, use a single-vehicle routing algorithm to form a route that will service the customers of that cluster (vehicle capacity constraints may be considered in this step of the procedure). The objective here is to minimize route length for each cluster. GCR offers four options with respect to which routing algorithm will be used in Step 3 (see Psaraftis (1980, 1983a, 1983b, 1983c)).

The main novelty (and core) of the GCR algorithm is its clustering step, whose basic purpose is to identify sets of customers that have good "directivity" characteristics, that is, have similar directional patterns of travel (e.g. "north to south", "east to west"). Such a task is very difficult in a many-to-many environment because customers whose origins are close to one another (and hence seem suitable to be picked up by the same vehicle) may have their destinations widely separated, thus making the eventual vehicle route inefficient. Step 2 of the algorithm was tested extensively and was shown to produce clusters of good "directivity".

2.6 Objective Function and Solution Method/ ADARTW. The solution method of ADARTW is completely different from that of GCR. In fact, ADARTW uses an explicitly defined objective function. The basic idea of the algorithm is to build routes and schedules by sequentially inserting each customer into the most promising "provisional" route and schedule constructed thus far. Its main novelty lies in the way by which feasible insertions are identified. The algorithm's basic version can be described roughly as follows (for complete details see Jaw, Odoni, Psaraftis and Wilson (1984)):

STEP 0 (Initialization): For all customers, set up time windows according to (3)-(8). Rank-order all customers by nondecreasing order of EPT. Put all customers into an "unprocessed customer list" (UCL). Let n = size of UCL. Set up "active vehicle fleet" (AVF) and "backup vehicle fleet" (BVF). Set up "pool size" K (K = user input ≥ 1)

STEP 1 (Candidate Customer Selection): Take the top k = min. (n, K) customers out of UCL and into "pool of customers for immediate assignment" (or, "pool"). If K = 0, END.

STEP 2 (Tentative Insertion): For all customers i in "pool", do the following:

STEP 2.1: For all vehicles j in AVF, do the following: Find all feasible ways in which i can be inserted into the work schedule of j so that (a) no service guarantees to any customers are violated; (b) no vehicle capacity constraint is violated (see also below). If none is found, examine another vehicle in AVF. If none exists, then either declare i REJECTED, or consider a vehicle j from BVF and repeat. Find the insertion of i into the work schedule of vehicle j which results in minimum marginal insertion cost (see also below). Call this cost $COST_{ij}$.

STEP 2.2: For customer i, find vehicle j*(i) for which $COST_{ij*(i)}$ $COST_{ij}$ for all j examined in Step 2.1. Call this minimum cost $MINCOST_i$. Tentatively assign i to j*(i).

STEP 3 (Best Insertion): Find customer i* in "pool" for which $MINCOST_i$ $MINCOST_i$ for all i in "pool". Permanently assign i* to j*(i*).

STEP 4 (Update): Update AVF and BVF. Update all vehicle schedules. Dump remaining customers of "pool" back into UCL. Set n = n-1 and go to Step 1.

Central to ADARTW is the concept of a "schedule block". A schedule block is a sequence of stops (pickups and deliveries) scheduled to be visited by a vehicle, with no idle time (or "slack") in between. Figure 3 shows two such schedule blocks: Block 1 represents the pickup and delivery of customers 1 and 2

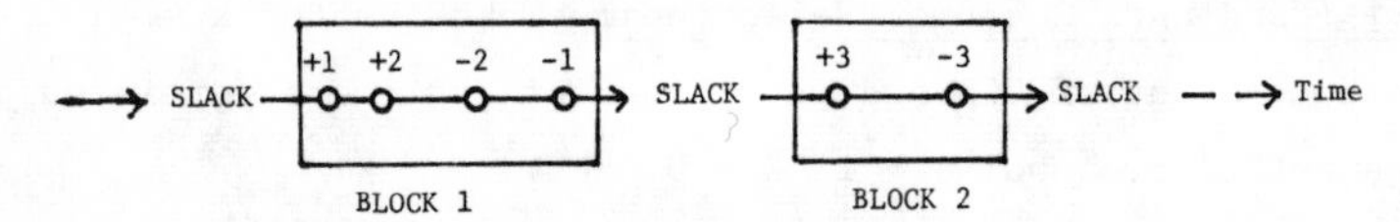

Figure 3: The concept of a schedule block (A "+" denotes pickup and a "-" denotes delivery)

(in the sequence shown) and Block 2 shows the pickup and delivery of customer 3. Slack time will generally exist between two consecutive schedule blocks. In Figure 3 a slack time between Blocks 1 and 2 means that the time it takes the vehicle to drive from the destination of customer 1 to the origin of customer 3 is less than the difference in the scheduled times of these two operations (deliver # 1 and pick up # 3). Such a schedule thus implies that the vehicle should idle at any one of these two points (or at both) for a total amount of time equal to the value of the slack.

Step 2.1 of ADARTW attempts to insert a new customer somewhere within the work-schedule of a vehicle. It does this by continually keeping track of how much each stop in each schedule block can be shifted upstream or downstream so that no constraints are violated. It also keeps track of the maximum available slack between consecutive blocks. Given a new customer and a candidate vehicle, there are three insertion options open to ADARTW: First, inserting both origin and destination within (or around) one individual schedule block. Second, creating a new schedule block by inserting these two points between two consecutive schedule blocks. And third, inserting them into two different schedule blocks. Notice that this last option implies that all slack between the two "receiving" blocks will have to disappear, since one of our operating assumptions is that the vehicle is not allowed to idle if a customer is on board (see Table 2). The basic mechanics of identifying feasible insertions for the first two options are described in Jaw, Odoni, Psaraftis and Wilson (1984)) and for all options in Jaw (1984). In this paper, we shall have an opportunity to observe the insertion logic of ADARTW in conjunction with the worst-case considerations of Section 5.

The marginal cost MC_{ij} of a (feasible) insertion of a customer i into the work schedule S of a vehicle j (with S already including other customers k) is defined as follows (subscript j dropped from all terms for convenience):

$$MC_i = DU_i + \sum_{k \epsilon S} (DU_k - DU_k) + VC_i \tag{12}$$

where DU_i (DU_k) is the disutility to customer i (k) <u>after</u> the insertion of customer i, DU_k is the disutility of customer k <u>before</u> the insertion of customer i and VC_i is the cost to vehicle resources.

The functional form assumed by ADARTW for DU_i (and, by analogy for DU_k and DU_k) is a quadratic function of x_i, the <u>pickup</u> or <u>delivery time deviation</u> of customer i, and, of y_i, the <u>excess ride time</u> of customer i, as follows:

$$DU_i = c_1 x_i + c_2 x_i^2 + c_3 y_i + c_4 y_i^2 \tag{13}$$

where x_i and y_i are defined as follows:

$$x_i = \begin{cases} APT_i - EPT_i & \text{if customer i is Type-P} \\ LDT_i - ADT_i & \text{if customer i is Type-D} \end{cases} \tag{14}$$

$$y_i = ART_i - DRT_i \tag{15}$$

and c_1 to c_4 are user-specified nonnegative constants.

A quadratic form was chosen for DU_i because it is both more general than a linear form and more realistic in representing actual customer dissatisfaction.

Finally, VC_i is assumed to be a function of z_i, the additional active vehicle time due to insertion of customer i, and of w_i, the additional slack vehicle time due to insertion of customer i, as follows:

$$VC_i = c_5 z_i + c_6 w_i + SW_i \, (c_7 z_i + c_8 w_i) \qquad (16)$$

where c_5 to c_8 are user-specified non-negative constants.

The term denoted by SW_i in (16) is called "system workload index" and is a measure of how heavy the workload of vehicle j is at the time of the insertion of customer i. The existence of this term in (16) penalizes additional vehicle time more in situations in which vehicle workload is already high. SW_i is calculated according to the following empirical formula:

$$SW_i = \frac{(\text{Number of system customers in } [EPT_i - T, EPT_i + T])}{(\text{Effective number of vehicles in above interval})} \qquad (17)$$

with T being another constant.

We note that in case i is not inserted into two different schedule blocks (which, in fact, would happen in most cases due to MRT-constraints and penalization of excess ride time y_i), the summation in (12) need only be evaluated for customers k belonging to at most one schedule block of S.

We end this section by noting that Table 2 displays some additional (but minor) features that help highlight the difference between GCR and ADARTW.

3. COMPUTATIONAL EXPERIENCE

There has already been extensive computational experience with both GCR and ADARTW. This has included running the algorithms with both simulated and real data, performing extensive sensitivity analyses on various parameters and, finally, comparing the two approaches on the same data set. The general assessment of these runs as far as both computational efficiency (CPU time) and performance (vehicle productivity, route circuity, customer service quality) are concerned has been quite positive for both approaches. In addition, both algorithms have dealt successfully with by far the largest-size problem database that has yet been examined (more than 2,600 customers), and can deal with even larger problems with no significant difficulty. This section summarizes these results and describes the cases on which the two algorithms were compared.

3.1 GCR Algorithm. In Jaw, Odoni, Psaraftis and Wilson (1982) a series of simulation runs of the GCR algorithm were described. These involved simulating 4 hours of service in a 6x6 square mile geographical area. Customer origins and destinations were distributed uniformly and independently over the area, and their requests were distributed over the 4-hour interval at a rate ranging from 100 to 500 customers per hour. A sensitivity analysis was made on the number of available vehicles (which ranged from 10 to 60), on the length of the time interval DT (which was chosen to be 30 or 60 minutes) and on the conversion factor CF (chosen to be 1.0 or 1.5). A 50%-50% mix of type-P and type-D customers was assumed. A general conclusion from these runs was that DT is the most important calibration parameter. In the runs examined, the best results were achieved with DT = 60 minutes whereas the performance of the algorithm generally deteriorated with DT smaller. This is probably due to the fact that GCR is less successful in linking adjacent time groups if DT is too small. Average vehicle productivity ranged from about 3.5 to more than 15 passengers per vehicle hour, depending on the scenario. Increasing the number of vehicles predictably resulted in lower productivity but in better quality of service, but did not necessarily result in higher CPU time, due to the tradeoff between the clustering and routing steps. This tradeoff is due to the fact that a higher number of vehicles generally results in more clusters (and hence more CPU time for the clustering step) but also in fewer customers per cluster (and hence less CPU time for the routing step). In all runs, CPU time ranged from 0.50 to about 16 minutes on a VAX 11/750, again depending on the size of the problem.

3.2 ADARTW Algorithm. Similar (although smaller-scale) simulation runs were performed in Jaw, Odoni, Psaraftis and Wilson (1984)) and in Jaw (1984) for ADARTW. A 9-hour, 6x6 square mile scenario was examined in Jaw, Odoni, Psaraftis and Wilson (1984), under the same (as in Section 3.1) general assumptions regarding the distribution of origins, destinations and desired times. In all cases the total number of customers were 250, with the demand pattern throughout the period of service fluctuating between 20 and 40 requests per hour. Time windows were set to 20 or 10 minutes and the maximum ride time was set to 5 minutes plus twice the direct ride time of each customer. An initial fleet size of 4 vehicles and an option to add more vehicles if necessary was assumed. Sensitivity analysis was performed on the values of the eight objective function coefficients and on the width of the time window. A general observation from these runs was that vehicle productivity (which ranged between 3 and 4 passengers per vehicle hour) seemed to be a decreasing function of c_1, c_2, c_3, and c_4, especially c_1 (the coefficient of the linear term representing pickup or delivery deviation). Average deviation from desired (pickup or delivery) time seemed to be most sensitive to and negatively correlated with both c_1 and c_2 (the coefficient of the quadratic term associated with pickup or delivery deviation), especially in the case of a wider time window. We should emphasize however that all of the above dependencies, although definitely identifiable, were not that pronounced. Thus, it seems (at least from these runs) that although the selection range of ADARTW's objective function coefficients is extremely broad, vehicle productivity and service attributes seem to depend less on the values of those coefficients, and more on other, problem-specific inputs, such as the time window width W, and the relationship between MRT and DRT (coefficients a and b). In that respect, a "tightly constrained" problem (W small, MRT close to DRT) predictably results in a lower productivity and in a higher number of vehicles if no customers are to be rejected. In our runs, between 10 and 14 vehicles were eventually used, this number never being a strictly monotonic function of any of the objective function coefficients. CPU time averaged about 20 seconds on the VAX 11/750.

It should be also stressed that all the runs of ADARTW reported in Jaw , Odoni, Psaraftis and Wilson (1984) referred to the version of the algorithm in which (a) the "pool size" K was set equal to 1 and (b) a customer's origin and destination could not be inserted into different schedule blocks. Larger pool sizes and the possibility of insertion into two different blocks were examined in detail in Jaw (1984). This analysis showed no significant improvement in productivity and service attributes for larger values of K, although this certainly resulted in a significant increase in CPU time. Nevertheless, we feel that a "small" value of $K \geq 1$ (say, between 5 and 10) is appropriate so as to reduce the myopia of ADARTW when $K = 1$ (an example of this is presented in Section 4). An insertion into different schedule blocks was observed to happen very rarely (for reasons that were discussed in Section 2.6) and to generally result in very small gains.

3.3 Comparison Between GCR and ADARTW. Despite the fact that the abstract problems (or, models) for which GCR and ADARTW have been designed are different, a comparison between the two approaches is indeed worthwhile, because both eventually refer to the same real-world problem. However, significant caution must be exercised in the design of the test cases so that the comparison is fair.

We have compared the two algorithms via two cases: one using real data and one using simulated data. Before we describe these cases, we observe that since GCR is "less constrained" than ADARTW (as far as quality of service, adherence to time constraints, etc. are concerned), one would a priori expect GCR to produce solutions of higher productivity than ADARTW if both procedures were applied to the same problem. Interestingly enough, the first comparison of the two algorithms produced exactly the opposite result. In that comparison ADARTW was observed to dominate GCR both in terms of quality of service and productivity. However, the second comparison revealed that such superiority of ADARTW is not always guaranteed, in the sense that GCR may, in some cases, achieve a better productivity than ADARTW without a significant deterioration in the quality of service. Details of the two comparisons are as follows:

The first comparison involved running both algorithms with a real database. The database covered about 16 hours of operation of the flexible-route system of Rufbus GmbH Bodenseekreis, in the city of Friedrichshafen, West Germany. The database included information on 2,617 customers, of which 2,397 were type-P while the remaining 220 were type-D (this database is maintained at M.I.T.).

Running either algorithm with this database presented some initial obstacles, because the database had features that did not match the operating assumptions of the procedures. Thus, some effort was spent to "convert" the original database to one that could be processed by both algorithms. Perhaps the most drastic of those conversions was the treatment of those type-P customers in the database who were "immediate-request" customers. For both algorithms, we converted these customers to "advance-request" customers by defining their DPT as the actual time of their request. In addition, since the distance matrix of the database was not symmetric (although asymmetries were not that pronounced), we ran the GCR algorithm with a converted distance matrix, by replacing each entry d_{ij} of the matrix with $0.5(d_{ij} + d_{ji})$. ADARTW was tested against the original matrix.

We were told that the Rufbus fleet consisted of one 33-passenger vehicle, four 9-passenger vehicles and twenty-three 17-passenger vehicles. Since both GCR and ADARTW assume identical vehicles, we ran ADARTW with a capacity of 17 for all vehicles. An "active vehicle fleet" of 10 vehicles, with the option to add more vehicles if necessary was assumed. In GCR (which does not accept capacity constraints) we externally varied the number of vehicles throughout the day with a peak of 21 vehicles. Vehicle speed was set to 15 mph for both procedures.

Finally, we were told that Rufbus schedulers tried to keep a 15-minute time window, although at times this constraint was relaxed to 60 minutes to avoid rejecting "immediate-request" customers. We somewhat arbitrarily translated this policy to the following parameters: For GCR, we set DT = 30 minutes and CF = 1.0. For ADARTW, we set W = 15 minutes, and MRT = 5 mins + 1.5*DRT. We selected the objective function coefficients of ADARTW as follows: $c_1=3$, $c_7=1$, $c_8=0.8$, and all others zero. Finally, we set the parameter T in (17) equal to 60 minutes.

The above description highlights some of the difficulties involved in testing any generic dial-a-ride algorithm on a real-life database, and the risks of comparing two procedures which have been based on different assumptions. Given the nature of some of the conversions, we have reservations about comparing _actual_ Rufbus schedules (which refer to a "mixed" demand case) with the results of GCR or ADARTW. However, we are more confident as far as the legitimacy of the comparison between GCR and ADARTW on this "converted" database is concerned. This comparison is displayed in Table 3 (Rufbus scheduling results are also displayed, for illustration purposes only).

TABLE 3: Comparison between GCR and ADARTW on the Rufbus "converted" database. Rufbus results are shown for illustration purposes only. All times are in minutes. Productivity is in passengers per vehicle hour. By convention, early deliveries (pickups) and late pickups (deliveries) have time deviations greater (less) than or equal to zero respectively.

	GCR	ADARTW	RUFBUS
Vehicles used	21	17	28
Vehicle productivity	10.53	12.06	8.87
Average deviation (late pickup or early delivery)	17.59	6.60	11.9
Average deviation (early pickup or late delivery)	-13.31	0	N.A.
Average ride time ratio	2.3	1.54	N.A.
CPU time (VAX 11/750)	7	12	N.A

We observe that both algorithms produce reasonable schedules with regard to quality of service and productivity levels. We also observe that, for this example, and with the exception of CPU time, ADARTW clearly dominates GCR on every count, that is, with respect to both quality of service and productivity. This dominance is perhaps surprising in view of the fact that the problem solved by ADARTW is more constrained than the one solved by GCR.

The second comparison was made based on a simulated database of only 25 customers (see Table 4). Customer origins and destinations were uniformly and independently distributed on a

TABLE 4: Simulated database of 25 customers. All X and Y coordinates are in hundredths a mile. All desired pickup and delivery times are in hours and minutes.

Customer (type)	Origin Coord.		Dest. Coord.		DPT	DDT
	X	Y	X	Y		
1 (D)	180	437	30	582		7:42
2 (D)	380	288	442	198		7:37
3 (D)	412	490	488	404		7:45
4 (P)	27	413	169	516	7:02	
5 (D)	197	29	205	153		7:56
6 (P)	326	490	324	166	7:14	
7 (D)	62	543	182	14		8:33
8 (D)	163	77	254	293		8:10
9 (P)	394	389	29	144	7:21	
10 (P)	31	66	283	278	7:23	
11 (D)	377	420	332	97		8:33
12 (D)	528	50	285	98		8:39
13 (P)	465	406	408	144	7:48	
14 (P)	472	73	93	34	7:51	
15 (D)	212	161	183	474		8:53
16 (D)	243	106	589	436		9:11
17 (D)	253	569	4	547		8:55
18 (D)	89	427	375	128		9:19
19 (D)	448	111	243	298		9:10
20 (D)	362	57	577	337		9:17
21 (D)	333	104	102	251		9:13
22 (D)	352	483	428	259		9:16
23 (P)	491	537	238	584	8:28	
24 (D)	348	33	520	259		9:27
25 (P)	27	194	33	299	8:40	

6x6 square mile area. Eight of those customers were type-P and 17 were type-D. Desired pickup (or delivery) times were uniformly distributed within approximately three hours of service. We ran both algorithms with 4 vehicles and a vehicle speed of 15mph. In GCR we used DT=60 minutes and CF=1.0, and in ADARTW W=30 minutes and MRT=5 minutes + 2*DRT. Finally, the objective function coefficients of ADARTW were set to $c_7 = 1$, $c_8 = 0.8$, and all others zero. Table 5 shows the results of the comparison. We can see that, in this example, GCR outperforms ADARTW in terms of total vehicle time, average ride time ratio and vehicle productivity, without using more vehicles. A further examination of the schedules (see Jaw (1984)) reveals that GCR delivers only two type-D customers more than 30 minutes earlier than their DDT's and <u>all</u> type-P customers less than 30 minutes later than their DPT's. Furthermore, no actual ride time in the GCR schedule exceeds the maximum ride time constraints of ADARTW.

TABLE 5: Comparison between GCR and ADARTW on the simulated database. All times are in minutes. Productivity is in passengers per vehicle hour. By convention, early deliveries (pickups) and late pickups (deliveries) have time deviations greater (less) than or equal to zero respectively.

	GCR	ADARTW
Vehicles used	4	4
Total vehicle time	360	389
# of late deliveries (out of 17)	3	0
Average deviation given late delivery	-16.3	0
Average deviation given early delivery	16.6	15.6
# of early pickups (out of 8)	3	0
Average deviation given early pickup	-9	0
Average deviation given late pickup	19.2	19.4
Average ride time ratio	1.33	1.49
Vehicle productivity	4.17	3.86
CPU Time (VAX 11/750)	1.12	1.4

Of course, the last comparison can be criticized on the grounds that ADARTW might conceivably obtain better results by an appropriate choice in its objective function coefficients. However, given that such choice is sometimes not clear, we believe that the comparison has demonstrated the premise that GCR can outperform ADARTW in some cases.

Due to small scale (and conceivably contrived nature) of this last comparison, it is clear that additional computational experience with both procedures is necessary in order to shed more light on what types of instances are likely to be more "favorable" to one of the algorithms as opposed to the other. However, based on our experience with both procedures to date, and on the fact that ADARTW possesses more features that would fit a given real-world situation (such as capacity constraints, general distance matrix, etc.), we conjecture ADARTW is likely to outperform GCR in most realistic operational situations and hence, is better suited for implementation in such situations. Still, we regard GCR as a viable tool for quickly producing reasonable solutions which could be used either in planning situations or in operational situations on a preliminary basis. In the latter case, GCR could be used to provide good starting solutions (e.g. customer clusters) that could be further improved with the help of a post-optimization routine (for instance, similar in spirit with the "swapper" algorithm of Bodin and Sexton (1982)).

The GCR and ADARTW codes have been written in FORTRAN IV and PL1 respectively, and their approximate lengths are 3000 lines for GCR and 2000 lines for ADARTW. These codes are maintained at M.I.T. (inquiries should be addressed to the author).

4. WORST-CASE CONSIDERATIONS IN ADARTW

Worst-case analyses of insertion heuritics have already been performed for several routing problems to-date. For instance, in

Rosenkrantz, Sterns and Lewis (1974), the worst-case error ratio of both the nearest insertion and the cheapest insertion methods, as applied to a "triangle-inequality" Traveling Salesman Problem of n nodes, was shown to be equal to 1.0. The worst-case error ratio of a heuristic H as applied to a minimization problem P is defined as the maximum, over all possible instances of P, values of the ratio $(Z_H - Z_{OPT})/Z_{OPT}$ where Z_H is the value of the objective Z of the problem when solved by H, and Z_{OPT} is its optimal value. (Such a definition implies that the two methods mentioned above produce tours which are at most 100% longer than the optimal tour, since in this case $(Z_H - Z_{OPT})/Z_{OPT} = 1.0$). For the same version of the Traveling Salesman Problem it is also known that both the arbitrary insertion and the farthest insertion methods exhibit a worst-case error ratio of 2*ln(n) - 0.84 (see also Golden, Bodin, Doyle and Stewart (1980)). More recently, a worst-case analysis of some heuristics for the vehicle routing and scheduling problem with time windows was carried out by Solomon (1983). This analysis showed that a variety of heuristics (including, but not limited to, insertion methods) exhibit an $\Omega(n)$ worst-case error ratio for n customers as far as the number of vehicles used, the total distance traveled, and the total schedule time are concerned (that is, the worst-case error ratio of these heuristics was shown to be at least equal to a linear function of n).

The pertinent question here is, whether any kind of similar analysis can be carried out for ADARTW (this issue is "academic" for GCR, since that approach does not use an explicitly defined objective function). A first observation is that such a task would be extremely difficult, given the nontrivial nature of the problem (regarding both its objective function and its constraints) and the nontrivial design of the algorithm. Jaw (1984) posed the question whether the $\Omega(n)$ behavior reported in Solomon (1983) might be true also for ADARTW, but left the question open.

Another observation that can be made is that analyzing heuristics in terms of their worst-case <u>relative</u> errors (that is, in terms of their worst-case error <u>ratios</u>) may make little sense in certain problems: for a minimization problem P and a heuristic H, if there are instances of P for which $Z_{OPT} = 0$ and Z_H 0, the worst-case error ratio of H becomes $+\infty$. This may lead to erroneous conclusions regarding the merits of H, particularly if the absolute error $Z_H - Z_{OPT}$ is small.

It is important to recognize that whereas such a "singular" ($Z_{OPT}=0$) behavior is rare in most vehicle routing problems (in which typically Z measures such quantities as total distance traveled, number of vehicles, etc., and hence $Z_{OPT} > 0$), this is not necessarily the case for the problem solved by ADARTW, due to the form of the objective function considered. Indeed, by an appropriate choice in the coefficients c_1 to c_8 (which are parts of the input that defines a specific problem instance), <u>it is relatively easy to envision cases for which</u> $Z_{OPT} = 0$. Such cases would correspond to objective functions measuring total customer disutility <u>only</u> (at least one of c_1, c_2, c_3, c_4 being nonzero and $c_5 = c_6 = c_7 = c_8 = 0$), in which the minimum achievable value of Z might go all the way down to zero. This means that if there exists such a case ($Z_{OPT} = 0$) for which $Z_{ADARTW} > 0$, the worst-case error ratio of ADARTW is $+\infty$.

Not surprisingly, such a case can be constructed. Despite the fact that the use of the ratio $(Z_{ADARTW} - Z_{OPT})/Z_{OPT}$ becomes meaningless when $Z_{OPT} = 0$, we shall present one such case below, so as to increase our understanding of how ADARTW behaves, and to indicate when and why it may behave poorly. We also discuss some alternative worst-case criteria later in this section.

Consider three requests on the Euclidean plane, as shown in Figure 4. Assume there is only one vehicle (of unit speed), initially located at point 0. Assume also that two of the requests (customers 2 and 3) have specified desired (earliest) <u>pickup</u> times equal to $\sqrt{2}/2 = 0.71$ and $1 + \sqrt{2}/2 = 1.71$

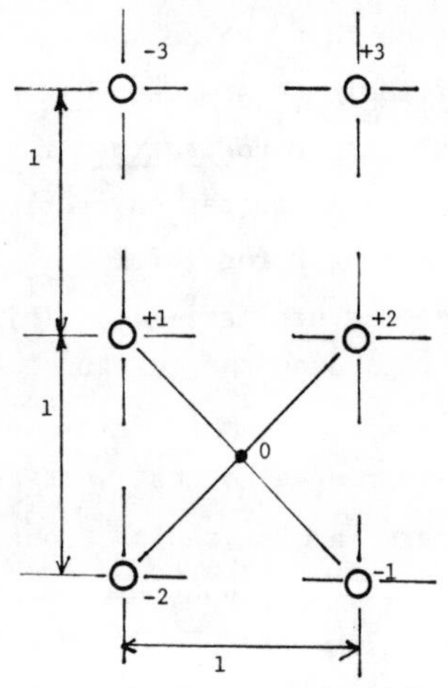

Figure 4: Customer and vehicle locations in example.

respectively. Customer 1 has specified a desired (latest) delivery time equal to $5 + \sqrt{2}/2 = 5.71$. Finally, assume that the time window W for this problem has been set equal to 6, and that the maximum ride time for all customers is equal to 4 (independent of direct ride time). It can be seen that the service guarantees for this problem are rather loose. Still, assume that we would like to obtain the lowest possible total customer disutility. For this problem, assume that the disutility of each customer is just equal to that customer's deviation from his desired time (pickup or delivery, as appropriate). That is, assume that the only nonzero objective function coefficient is $c_1 = 1$.

Given the above data, Table 6 displays the calculated

TABLE 6: Time windows for 3 customers of example (asterisked times are inputs; all other times are calculated).

Customer	Type	EPT	LPT	EDT	LDT
1	D	-4.29	4.29	-0.29	5.71*
2	P	0.71*	6.71	2.12	10.71
3	P	1.71*	7.71	2.71	11.71

(according to (3) - (8)) pickup and delivery time windows for all three customers.

A cursory investigation reveals that the optimal solution to this problem corresponds to the route (0, +2, +3, -3, +1, -2, -1) shown in Figure 5. The corresponding schedule is shown in Table 7. This solution is optimal because the total deviation from desired time is zero ($Z_{OPT} = 0$), the lowest possible. Notice also that there is no slack in the vehicle schedule.

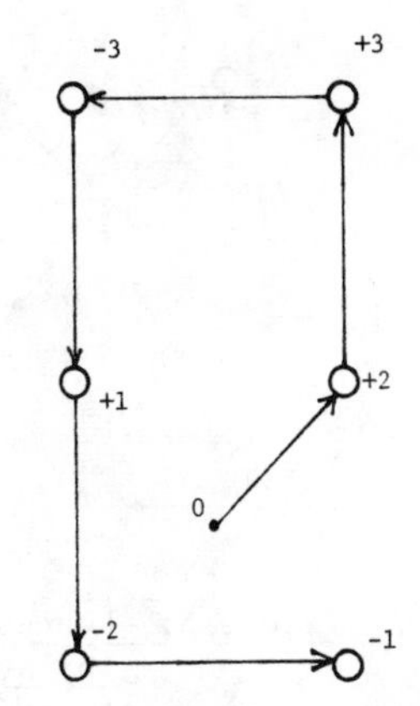

Figure 5: Optimal route (see also Table 7

Table 7: Optimal Schedule (see also Figure 5)

Stop	Arrival Time	Departure Time	Idle Time	Desired Time	Time Deviation	Ride Time
0	0	0	0	-	-	-
+2	0.71	0.71	0	0.71	0	-
+3	1.71	1.71	0	1.71	0	-
-3	2.71	2.71	0	-	-	1
+1	3.71	3.71	0	-	-	-
-2	4.71	4.71	0	-	-	4
-1	5.71	-	-	5.71	0	2

ADARTW solves the above problem as follows: If the "pool size" K is set equal to one, the procedure starts by first considering customer 1, the customer with the lowest EPT. Since he is a type - D customer, ADARTW schedules him/her in such a way that his/her delivery time deviation is as small as possible, that is, zero. The resulting route and schedule are shown in Figure 6a and Table 8a respectively. Note that there is a single schedule block ("Block 1") in the schedule, which is pushed as late in time as possible to achieve zero delivery time deviation for customer 1. This necessitates a slack period of 3.59, during which the vehicle idles at 0. Up to this iteration, $Z_{ADARTW} = 0$.

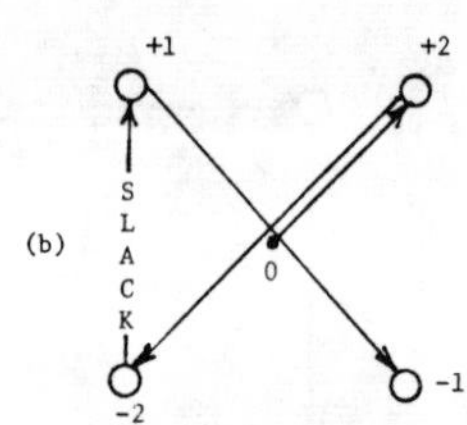

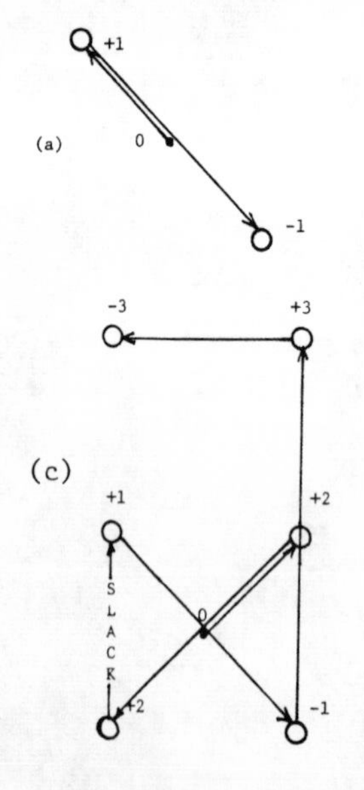

Figure 6: Routes for (a) first, (b) second, and (c) third iterations of ADARTW (see also Tables 8a, 8b, and 8c respectively).

TABLE 8a: Schedule at first iteration of ADARTW (also see Figure 6a)

Stop	Arrival Time	Departure Time	Idle Time	Desired Time	Time Deviation	Ride Time
0	0	3.59	3.59	-	-	-
+1	4.29	4.29	0	-	-	-
-1	5.71	-	-	5.71	0	1.41

ADARTW next considers customer 2. The best insertion results in a route and schedule shown in Figure 6b and Table 8b respectively. The schedule consists now of two blocks. Block 1 remains as before, that is, pushed as late in time as possible. The new block (which includes the pickup and delivery of customer 2) is Block 2, which is pushed as early in time as possible, to

TABLE 8b: Schedule at second iteration of ADARTW (also see Figure 6b)

Stop	Arrival Time	Departure Time	Idle Time	Desired Time	Time Deviation	Ride Time
0	0	0	0	-	-	-
+2	0.71	0.71	0	0.71	0	-
-2	2.12	3.29	1.17	-	-	1.41
+1	4.29	4.29	0	-	-	-
-1	5.71	-	-	5.71	0	1.41

achieve a zero pickup time deviation for customer 2 (who is of type P). There is a residual slack of 1.17 between these two blocks, representing the total amount of available idle time inbetween. As before, and up to this iteration, $Z_{ADARTW} = 0$.

Perhaps the most crucial observation at this point is that since the delivery of customer 2 already precedes the pickup of customer 1, there is no way that this relative order can be reversed in subsequent insertions. In other words, we can already see that it would be impossible for ADARTW to insert customer 3 so that it would eventually produce the optimal route and schedule of Figure 5 and Table 7. Still, the possibility of obtaining an alternate optimal route and schedule cannot be ruled out before all possible insertions of customer 3 are evaluated.

Unfortunately, there are not that many feasible insertions for customer 3, even if insertion into two different schedule blocks is attempted. Inserting both +3 and -3 into (including before or after) Block 2 is infeasible, primarily because customer 3's EPT is rather late (1.71) and because the slack between Blocks 1 and 2 is rather tight (1.17). The same is true regarding an insertion into two different blocks. The only possible insertion seems to be the one immediately after Block 1 (see Figure 6c and Table 8c). Such an insertion would entail a pickup time deviation of 6 for customer 3, if the delivery time deviation of customer 1 is to remain at zero. This proves that for this example $Z_{ADARTW} = 6$ while $Z_{OPT} = 0$, that is, ADARTW is not successful in identifying the optimal solution.

TABLE 8c: Schedule at third (and final) iteration of ADARTW (also see Figure 6c).

Stop	Arrival Time	Departure Time	Idle Time	Desired Time	Time Deviation	Ride Time
0	0	0	0	-	-	-
+2	0.71	0.71	0	0.71	0	-
-2	2.12	3.29	1.17	-	-	1.41
+1	4.29	4.29	0	-	-	-
-1	5.71	5.71	0	5.71	0	1.41
+3	7.71	7.71	0	1.71	6	-
-3	8.71	-	-	-	-	1

ADARTW fails to produce the optimal route and schedule in this example for a number of reasons, due both to certain features of the instance itself and to some features of the algorithm. Regarding the instance itself, one could possibly identify the combination of loose service guarantees (W = 6, MRT = 4), coupled with a mix of type-P and type-D customers as conducive to the type of behavior displayed above. Regarding the algorithm, it is clear that the fact that the "pool size" K was set equal to 1, and the fact that ADARTW has no "resequencing" capability ultimately forced the algorithm to insert customer 3 after Block 1.

It can be seen also that the above two features can cause ADARTW to mistakenly declare a feasible problem instance as infeasible. Indeed, it is easy to check that such would be the outcome in the example examined if the width of the time window W were equal to 4 instead of 6. In such a case, the optimal route and schedule would remain unchanged, but ADARTW would be unable to service customer 3 without an additional vehicle.

As a brief aside, one might wonder how would the GCR algorithm perform on such a "perverse" instance. That would depend, to a significant extent, on the values of the parameters DT and CF of that procedure. It is straightforward to check that if CF = 1.5, and if time groups of DT = 2 are set up starting at t = 0, the GCR algorithm would produce the optimal route and schedule for this example. Of course, other choices in DT and CF might make GCR fail to identify the optimal solution (an example for this case is DT = 3, CF = 1.0).

This analysis has been to a certain extent unfair to ADARTW, because the criterion that has been used (worst-case error ratio) is less than perfectly suited for this problem. One could actually imagine even more extreme cases, such as a 1,000-customer problem in which $Z_{OPT} = 0$ and in which ADARTW achieves a zero time deviation for 999 customers and a time deviation of 1 for one customer. It would be clearly nonsensical to state that for this case ADARTW performs infinitely poorly. In addition, the analysis is unfair because it refers to pathological cases which are unlikely to occur in the real world. Indeed, it is not uncommon for a heuristic whose worst-case performance is poor to behave decently in practice; perhaps the most typical example in this context is the k-interchange heuristic of Lin and Kernighan (1973) for the TSP, which, in spite of an arbitrarily poor worst-case behavior (Papadimitriou and Steiglitz (1978)), is known to be one of the best heuristics devised for the TSP to date. Our real-world computational experience with ADARTW, which has been very encouraging thus far, tends to confirm the disparity between worst-case and average-case behavior of the algorithm.

As far as the worst-case criterion used is concerned, we note here that there is considerable room for further analysis if ADARTW's performance is evaluated otherwise. Error ratios can still be used if the objective function coefficients are restricted to certain values that would guarantee $Z_{OPT} > 0$. For instance, an interesting question is what happens, in terms of worst-case performance, if the only nonzero coefficient is c_5. That is, the issue is what is ADARTW's worst-case performance in terms of total vehicle active time (or, equivalently, total vehicle distance traveled). The worst that could happen here might be a case in which ADARTW is forced to use n vehicles for n customers traveling from the same origin to the same destination, whereas the optimal solution recommends only one vehicle for these customers. Thus far, we have been unable to construct such a pathological instance if the only nonzero coefficient is c_5.

Error ratios can also be used if they compare the heuristic error to the maximum possible error rather than the optimal value of Z. A number of researchers (see for instance Fisher (1980)) have suggested the use of the ratio $(Z_H - Z_{OPT})/(Z_R - Z_{OPT})$ in lieu of the traditional error ratio $(Z_H - Z_{OPT}/Z_{OPT})$ for minimization problems, where Z_R is a suitably chosen reference value. Ideally, Z_R should be the maximum value of Z. Unfortunately, this maximum value is usually no easier to identify than Z_{OPT}, so Z_R is instead taken to be some upper bound on the maximum value of Z that is presumably easier to obtain. In terms of the previous example ($c_1 = 1$ and all others zero), it can be seen that the maximum possible total time deviation is equal to the time window width multiplied by the total number of customers, that is, $Z_R = nW$ (in this case $Z_R = 18$). If this is the case, the solution error $Z_{ADARTW} - Z_{OPT} = 6$ seems quite reasonable as compared to the maximum possible error $Z_R - Z_{OPT} = 18$.

A final direction in the worst-case analysis of ADARTW is to abandon the use of error ratios altogether and focus on absolute errors instead (that is, use $Z_{ADARTW} - Z_{OPT}$ as the error criterion). The presence of hard constraints implies that upper bounds on the total customer deviation from desired (pickup or delivery) time, on the total excess ride time, and on the total vehicle active time are equal to nW, $\sum_{i=1}^{n} (MRT_i - DRT_i)$ and $\sum_{i=1}^{n} (D(depot, +i) + MRT_i)$ respectively, if no customers are to be rejected. However, these bounds are likely to be loose because they fail to take into account the nature of the algorithm and other problem inputs (such as vehicle capacity). Whether these (and possibly other) bounds can be tightened by exploiting the special structure of the problem and the algorithm remains to be seen.

5. CONCLUDING REMARKS

This paper has reviewed two heuristic algorithms developed by the author and his colleagues for the multi-vehicle advance-request Dial-A-Ride problem, and has summarized computational experience with the procedures to date. Both algorithms seem to be quite efficient computationally, and have solved the largest (to our knowledge) instances in this problem class to date. On the basis of this computational experience, we feel that both procedures can be useful in the implementation of an advance-request Dial-A-Ride system. We feel the GCR algorithm can be best used as a fast planning tool and/or as a device to produce good starting solutions in an operational situation, whereas the ADARTW algorithm can form the basis of an operational scheduling system that would assist the dispatcher in the actual execution of the schedule. Of course, additional refinements in both procedures, and continuing computational experience with them are necessary to both shed more light on their performance, and, ultimately, enhance that performance even more.

We end this paper by discussing several important issues on the scenarios under which the procedures of the paper are likely to be implemented.

We mentioned earlier that the issue of whether customer service guarantees (or, the corresponding time constraints) should be "hard" or "soft" in an advance-request environment is a debatable one. The "hard" approach is certainly more appealing from a policy (or even a public relations) standpoint, plus, lends itself more easily to analysis by quantitative techniques. In addition, there are certainly cases in which a customer has to be picked up (or delivered) within a prescribed time window. However, the "hard" approach also has some drawbacks: First, it opens the door to "infeasible solutions" or to solutions which are erroneously declared as infeasible (as seen in Section 4). How would a dispatcher handle an "infeasible solution" in practice? Rejecting customers may not be an allowable alternative. Similarly, adding more vehicles to make the problem feasible may be far less implementable an option than the discussion thus far would seem to suggest. This is certainly true in a "mixed" demand scenario where the dispatcher might simply be unable to add vehicles upon request. Thus, it may happen that in such situations the best alternative for the dispatcher would be to make the solution feasible by relaxing the constraints. This could possibly involve rerunning the algorithm, and/or calling some or all of the customers to renegotiate new service parameters. Leaving aside for the moment the fact that a "hard" constraint that is subsequently relaxed is, by definition, "soft", one can also see that there are a multitude of other public perception problems that could occur under these (or similar) circumstances. This is particularly true if the (supposedly "hard") constraints under which the algorithm operates have been advertised as such in public.

Extending an advance-request dial-a-ride algorithm into the equivalent "mixed" demand case is a well-motivated task, since pure advance-request systems are either nonexistent or very rare. Such an extension would be much more difficult to implement in the GCR algorithm than in ADARTW. Indeed, the design of ADARTW would make the real-time insertion of immediate-request customers into an advance-request schedule obtained (say) the previous day, seem "straightforward". In fact, certain facets of the insertion problem become easier for immediate-request customers (for instance, schedules can no longer be shifted <u>earlier</u> in time upon appearance of an immediate request, and that would alleviate some of the computational burden of the procedure). However, as alluded to above, there are a number of issues that merit serious attention before such an extension is implemented. Given that it might be very difficult, or even impossible, to add new vehicles so as to make the problem feasible each time an immediate request appears, one would likely have to adopt a different system of service guarantees for real-time requests. In fact, it is exactly because of such issues that Rufbus policy makers decided to consider time constraints as "soft" rather than "hard" (see the description in Section 3.3). Short of scrapping the idea of "hard" time constraints altogether, it might make sense to use it for advance-request customers only, and adopt softer service guarantees for real-time customers.

Finally, there are a number of other ideas that can be implemented to further enhance ADARTW. For instance, one could easily modify the algorithm to account for nonzero dwell times, or for specialized vehicles (for special categories of customers such as wheelchair customers, etc.). One could also easily add a post-optimization module such as k-interchange to provide a "resequencing" capability to the algorithm, or even a capability to "swap" customers among vehicles. Whether the last two measures would <u>substantially</u> enhance the performance of ADARTW is an open question at this point.

ACKNOWLEDGEMENTS

This work was supported in part by the Paratransit Integration Program of the Urban Mass Transportation Administration, U.S. Department of Transportation. The assistance of Ed Neigut, the project's technical monitor, on the formulation of the problem solved by ADARTW is gratefully acknowledged. Sincere thanks are also due to Horst Gerland of Rufbus, Wolfgang Kratschmer of Dornier, and others in their team for making their database available, and for many fruitful discussions. Last, but not least, the author would like to thank his project colleagues, Jang-Jei Jaw, Amedeo Odoni, and Nigel Wilson, without whom this work would not have been completed, and Bruce Golden, an anonymous referee and the Editor for their comments.

REFERENCES

Bodin, L.D. and Sexton, T.R. (1982). The Multi-Vehicle Subscriber Dial-A-Ride Problem. Working Paper No. 82-005, Department of Management Science and Statistics, College of Business and Management, University of Maryland, College Park, Maryland 20742.

Fisher, M.L. (1980). Worst-Case Analysis of Heuristic Algorithms. Management Science 26, 1-16.

Golden, B., Bodin, L.D., Doyle, T., Stewart, W., Jr. (1980). Approximate Traveling Salesman Algorithms. Operations Research 28, 694-711.

Hung, H.K., Chapman, R.E., Hall, W.G., Neigut, E. (1982). A Heuristic Algorithm for Routing and Scheduling Dial-A-Ride Vehicles. ORSA/TIMS National Meeting, San Diego, California.

Jaw, J.J. (1984). Heuristic Algorithm for Multi-Vehicle, Advance-Request Dial-A-Ride Problems. Ph.D. Thesis, Department of Aeronautics and Astronautics, M.I.T., Cambridge, Massachusetts 02139.

Jaw, J.J., Odoni, A.R., Psaraftis, H.N., and Wilson, N.H.M. (1982). A Heuristic Algorithm for the Multi-Vehicle Many-to-Many Advance-Request Dial-A-Ride Problem. Working Paper MIT-UMTA-82-3, M.I.T., Cambridge, Massachusetts 02139.

Jaw, J.J., Odoni, A.R., Psaraftis, H.N., and Wilson, N.H.M. (1984). A Heuristic Algorithm for the Multi-Vehicle Advance-Request Dial-A-Ride Problem with Time Windows. Transportation Research (8), to appear.

Lin, S., Kernighan, B.W. (1973). An Effective Heuristic Algorithm for the Traveling Salesman Problem. Operations Research 21, 498-516.

Papadimitriou, C.H., Steiglitz, K. (1978). Some Examples of Difficult Traveling Salesman Problems. Operations Research 26, 434-443.

Psaraftis, H.N. (1980). A Dynamic Programming Solution to the Single Vehicle Many-to-Many Immediate Request Dial-A-Ride Problem. Transportation Science 14, 130-154.

Psaraftis, H.N. (1983a). An Exact Algorithm for the Single Vehicle Many-to-Many Dial-A-Ride Problem with Time Windows. Transportation Science 17, 351-357.

Psaraftis, H.N. (1983b). k-Interchange Procedures for Local Search in a Precedence-Constrained Routing Problem. European Journal of Operational Research 13, 391-402.

Psaraftis, H.N. (1983c). Analysis of an $O(N^2)$ Heuristic for the Single Vehicle Many-to-Many Euclidean Dial-A-Ride Problem. Transportation Research 17B, 133-145.

Rosenkrantz, D.J., Stearns, R.E., Lewis, P.M. (1974). Approximate Algorithms for the Traveling Salesperson Problem. Proceedings of the 15th Annual IEEE Symposium of Switching and Automata Theory, 33-42.

Roy, S., Chapleau, L., Ferland, J., Lapalme, G., and Rousseau, J.M. (1983). The Construction of Routes and Schedules for the Transportation of the Handicapped. Working Paper, Publication No. 308, Centre de Recherche sur les Transports, University of Montreal, Case Postale 6128 - Succursale A, Montreal, PQ, Canada H3C 3J7.

Sexton, T.R. (1979). The Single Vehicle Many-to-Many Routing and Scheduling Problem. Ph.D. Thesis, State University of New York at Stony Brook, New York 11738.

Solomon, M.M. (1983). On the Worst-Case Performance of Some Heuristics for the Vehicle Routing and Scheduling Problem with Time Window Constraints. Working Paper 84-02, Graduate School of Business Administration, Northeastern University, Boston, Massachusetts 02115.

Wilson, N.H.M., Weissberg, H. (1976). Advanced Dial-A-Ride Algorithms Research Project: Final Report. Report R76-20. Department of Civil Engineering, M.I.T. Report R-76-20, Cambridge, Massachusetts 02139.

Wilson, N.H.M., Colvin, N.H. (1977). Computer Control of the Rochester Dial-A-Ride System. Report R77-31. Department of Civil Engineering, M.I.T. Report R77-31, Cambridge, Massachusetts 02139.

Received 11/84; Revised 8/25/86.

AMERICAN JOURNAL OF MATHEMATICAL AND MANAGEMENT SCIENCES

PICKUP AND DELIVERY OF PARTIAL LOADS WITH "SOFT" TIME WINDOWS

Thomas R. Sexton
W. Averell Harriman College for Policy Analysis and Public Management
State University of New York
Stony Brook, New York 11794

Young-Myung Choi
Korea Advanced Energy Research Institute
P.O. Box 7, Daeduk-Danji,
Choong-Nam, Korea 300-31

SYNOPTIC ABSTRACT

We apply Benders' decomposition procedure to the single-vehicle routing and scheduling problem with time windows, partial loads, and dwell times. We provide a formulation and demonstrate that the scheduling subproblem is the dual of a network flow problem. We describe an exact algorithm which exploits its structure, and construct a route improvement heuristic based on the master problem. A heuristic for building an initial route is also presented.

Key Words and Phrases: vehicle routing and scheduling; time windows; Benders' decomposition.

1986, VOL. 6, NOS. 3 & 4, 369-398
0196-6324/86/030369-30 $9.00

1. INTRODUCTION.

In this paper, we consider a problem called the single-vehicle routing and scheduling problem with time windows, partial loads, and dwell times. It is a many-to-many problem, meaning that each load has its own origin and its own destination (there is no depot). We assume that the vehicle can be made available at any pickup site at any time to start the route. Each load occupies a known fraction of the vehicle, which is called its "load factor." Associated with the pickup and delivery of each load are time intervals called pickup and delivery time windows, and dwell times (which are the amounts of time required to load and unload at each origin and destination). The problem involves finding a sequence of pickups and deliveries (the route) which satisfies the capacity restriction of the vehicle as well as obvious precedence requirements, and a set of times at which these tasks are to be performed (the schedule) so as to minimize a linear combination of (1) total vehicle operating time, and (2) total customer penalty due to missing any of the time windows.

Specifically, total vehicle operating time is defined as the time from the beginning of the first pickup until the completion of the last delivery. The pickup time window deviation for a load is zero if the pickup of the load begins within its specified interval, and equals the absolute difference between the beginning of the pickup and the nearest endpoint of the interval otherwise. The delivery time window deviation for a load is defined similarly using the time at which the delivery of the load begins. The penalty associated with a load is a weighted sum of the pickup and delivery window deviations and the total customer penalty is defined as the sum over all loads of the individual load penalties.

We call this the "soft window" version of this problem because we do not require that the tasks begin within the given time windows, like the "hard window" version does, but rather penalize the solution whenever a time window is missed. We believe that the soft window version is more realistic because (1) it always has a feasible solution, and (2) the windows can be made arbitrarily hard by increasing the coefficients used to compute the load penalties relative to that which multiplies total vehicle operating time. Thus the hard window problem may be viewed as the limit of the soft window problem as certain coefficients (e_i^1 and e_i^2 in Section 2 below) go to infinity. This approach permits considerable flexibility in actual applications and avoids the useless "no feasible solution" situation which is possible in the hard window problem.

Excellent overviews of the current state-of-the-art in routing and scheduling can be found in Bodin, Golden, Assad, and Ball (1983), Bodin and Golden (1981) and Magnanti (1981). Lenstra and Rinnooy Kan (1981) showed that most routing and scheduling problems are NP-hard, which means that they belong to a class of problems for which no polynomial time algorithm is known. Consequently, it is not surprising that success with exact procedures has been limited.

Solomon (1978) used a formulation first presented by Fisher and Jaikumar (1981) to solve the vehicle routing and scheduling problem with "hard" time windows. This problem calls for the vehicle to visit each site within its time window, pick up its load, and then return all collected loads to a central depot. In reverse, the problem calls for the delivery of loads to the sites from the central depot. In either case, however, this problem is free of the precedence constraints in our problem, which require that each load be picked up at its origin before being delivered at its destination.

Ball, Golden, Assad, and Bodin (1983) and Love (1981) considered the trailer routing problem with full loads. Each devised a heuristic algorithm in a mathematical formulation. The full load problem is considerably less complex than the partial load problem since feasible routes must be alternating sequences of pickups and deliveries.

Assad, Ball, Bodin, and Golden (1981) developed a computer-based system for a large commercial firm which employed a heuristic algorithm for the multi-vehicle partial load problem. Sexton (1979) and Sexton and Bodin (1985a, 1985b, 1981) used Benders' decomposition of a mixed integer formulation to approximately solve the related single vehicle dial-a-ride problem and then embedded the procedure in a multi-vehicle environment in Bodin and Sexton (to appear). Stein (1978) conducted an asymptotic probabilistic analysis of the multi-vehicle dial-a-ride problem and Cullen, Jarvis and Ratliff (1981) used set partitioning in an interactive setting on a version of the same problem. Armstrong and Garfinkel (1982) used dynamic programming to solve the dial-a-ride problem with time windows exactly, as did Psaraftis (1980, 1983). These procedures are computationally reasonable for problem instances involving fewer than about ten customers.

The formulation and algorithm we present in the present paper are very similar to those used by Sexton (1979), Sexton and Bodin (1985a, 1985b, 1981) and Bodin and Sexton (to appear) for the single and multiple vehicle dial-a-ride problems. The single vehicle dial-a-ride problem requires that customers be picked up and delivered at individual locations so as to minimize a linear combination of total customer excess ride time and total customer delivery time deviation. A customer's excess ride time is the additional time he spends on the vehicle beyond the minimum ride time from his origin to his destination while his delivery time deviation is the difference between his actual and his prestated desired delivery time.

The capacity and scheduling constraints of the dial-a-ride problem are very similar to those presented below but the dial-a-ride problem, as stated in the references cited, has no time windows. Its objective function is thus quite different and its scheduling network is simpler than that presented below. The dial-a-ride scheduling network is acyclic so that negative cost cycles can never arise, while in the present problem we must identify and saturate all such cycles in order to ensure optimality. The routing procedure, on the other hand, is virtually identical in both the dial-a-ride and the present algorithm. This is true of both the initial routing heuristic and the route improvement procedure described below.

2. FORMULATION AND ANALYTICAL APPROACH.

Let n be the number of loads to be picked up and delivered. Denote by $[\ell_i^0, u_i^0]$ and $[\ell_i^1, u_i^1]$ the pickup and delivery time windows, respectively, of load i. These are the time intervals within which the corresponding tasks must begin. Also, let x_i^0 and y_i^0 denote the midpoints of the pickup and delivery time windows, respectively. Next, let δ_i^0 and δ_i^1 be the pickup and delivery dwell times for load i. These are the times required by the vehicle to effect the loading and unloading at the origin and destination of load i.

Let $d_{00}(i,j)$, $d_{01}(i,j)$, $d_{10}(i,j)$ and $d_{11}(i,j)$ be the travel times among the locations. We have adopted the following notation convention: the first subscript represents the starting location of the trip and the second subscript its ending location; a subscript equal to zero represents an origin and a subscript equal to one represents a destination; the arguments represent the load numbers at the beginning and at the end of the trip, respectively. Thus $d_{10}(6,3)$ is the travel time from the destination of load 6 to the origin of load 3. We assume that travel times are symmetric, namely $d_{00}(i,j)=d_{00}(j,i)$, $d_{01}(i,j)=d_{10}(j,i)$ and $d_{11}(i,j)=d_{11}(j,i)$, and that the triangle inequality holds, i.e., $d_{pq}(i,j)\leq d_{ps}(i,k)+d_{sq}(k,j)$ for all i,j=1,...,n and p,q,s= 0,1. In addition, we define the following modified travel times:

$$\begin{cases} d'_{00}(i,j) = d_{00}(i,j) + \delta_i^0 \\ d'_{01}(i,j) = d_{01}(i,j) + \delta_i^0 \\ d'_{10}(i,j) = d_{10}(i,j) + \delta_i^1 \\ d'_{11}(i,j) = d_{11}(i,j) + \delta_i^1, \end{cases}$$

which represent <u>the minimum time from the beginning of the first task on a leg of the route to the beginning of the second task on that leg.</u>

There are two types of variables in this problem: routing variables that indicate the <u>sequence</u> in which the tasks are to be performed, and scheduling variables that specify the <u>times</u> at which they are to be performed. <u>The routing variables</u> in this formulation are

$$u_{ij} = \begin{cases} 1, & \text{if the pickup of load i precedes that of load j} \\ 0, & \text{otherwise} \end{cases}$$

$$v_{ij} = \begin{cases} 1, & \text{if the pickup of load i precedes the delivery of load j} \\ 0, & \text{otherwise} \end{cases}$$

$$w_{ij} = \begin{cases} 1, & \text{if the delivery of load i precedes that of load j} \\ 0, & \text{otherwise} \end{cases}$$

for $i,j=1,2,\ldots,n$, $i \neq j$. <u>The scheduling variables</u> are, for $i=1,2,\ldots,n$,

x_i = time at which the pickup of load i begins.

y_i = time at which the delivery of load i begins.

<u>The pickup and delivery time window deviations for load i</u> are

$$f_i(x_i) = \begin{cases} \ell_i^0 - x_i & \text{if } x_i < \ell_i^0 \\ 0 & \text{if } \ell_i^0 \leq x_i \leq u_i^0 \\ x_i - u_i^0 & \text{if } x_i > u_i^0 \end{cases}$$

$$g_i(y_i) = \begin{cases} \ell_i^1 - y_i & \text{if } y_i < \ell_i^1 \\ 0 & \text{if } \ell_i^1 \leq y_i \leq u_i^1 \\ y_i - u_i^1 & \text{if } y_i > u_i^1 \end{cases}$$

which can be expressed as

$$f_i(x_i) = (1/2)\ \{|x_i - x_i^0| - x_i^0 + \ell_i^0 + \left||x_i - x_i^0| - x_i^0 + \ell_i^0\right|\}$$

and

$$g_i(y_i) = (1/2)\ \{|y_i - y_i^0| - y_i^0 + \ell_i^1 + \left||y_i - y_i^0| - y_i^0 + \ell_i^1\right|\}$$

Total vehicle operating time is given by:

$$y_{(n)} + \delta_{(n)}^1 - x_{(1)},$$

where the subscripts (1) and (n) refer to the first and last tasks in the route sequence. Of course, we do not know yet which tasks these are but, as we will see, our solution procedure does not require this information.

Our model, then, is seeking to minimize

$$z' = y_{(n)} + \delta_{(n)}^1 - x_{(1)} + \sum_{i=1}^{n} [e_i^1 f_i(x_i) + e_i^2 g_i(y_i)] \qquad (1)$$

subject to

$$\sum_{\substack{j=1 \\ j \neq i}}^{n} a_j u_{ji} v_{ij} \leq 1 - a_i \quad i=1,2,\ldots,n \qquad (2)$$

$$y_i - x_i \geq d'_{01}(i,i) \quad i=1,2,\ldots,n \qquad (3)$$

$$d'_{00}(j,i) \leq x_i - x_j + Mu_{ij} \leq M - d'_{00}(i,j)\ ;\ i \neq j \qquad (4)$$

$$d'_{10}(j,i) \leq x_i - y_j + Mv_{ij} \leq M - d'_{01}(i,j)\ ;\ i \neq j \qquad (5)$$

$$d'_{11}(j,i) \leq y_i - y_j + Mw_{ij} \leq M - d'_{11}(i,j)\ ;\ i \neq j \qquad (6)$$

$$x_i\ ,\ y_i \text{ unrestricted in sign } \quad i = 1,2,\ldots,n \qquad (7)$$

$$u_{ij}\ ,\ v_{ij}\ ,\ w_{ij} = 0 \text{ or } 1 \quad i \neq j \qquad (8)$$

where e_i^1 and e_i^2 are the positive constants referred to earlier, i=1,2,...,n, and where M is an arbitrarily large positive constant discussed below. We observe that as e_i^1 and e_i^2 grow to infinity, the penalties associated with missing the windows also grow, meaning that the "soft" windows become "hard."

Constraint set (2) guarantees that the vehicle is never overloaded. This is strictly a routing constraint, and hence involves only routing variables. Observe that the vehicle can only become overloaded at a pickup location. From the definitions of the routing variables, load $j \neq i$ will be on the vehicle immediately following the pickup of load i if and only if $u_{ji} = v_{ij} = 1$. Thus the total load on the vehicle immediately following the pickup of load i, including load i itself, is

$$a_i + \sum_{\substack{j=1 \\ j\neq 1}}^{n} a_j u_{ji} v_{ij},$$

which must not exceed unity, the vehicle's capacity. Note that this argument holds even if loads are being picked up at the locations at which others are being delivered.

Constraint sets (3) through (6) are designed to ensure that the vehicle's schedule allows it sufficient time to travel between the locations specified by the route and to perform the necessary loading and unloading tasks. In particular, constraint set (3) guarantees that every pickup precedes its own delivery.

To illustrate the nature of constraint set (4), consider the pickup times of two distinct loads i and j. If $u_{ij} = 1$, then we must have

$$x_j - x_i \geq d'_{00}(i,j)$$

whereas if $u_{ij} = 0$, we must have

$$x_i - x_j \geq d'_{00}(j,i).$$

Constraint set (4) embodies both conditions provided M is much larger than any d' values. Constraint sets (5) and (6) follow by very similar arguments, and constraint sets (7) and (8) are strictly definitional.

The problem of minimizing (1) subject to (2) through (8) is called the single-vehicle routing and scheduling problem with time windows, partial loads, and dwell times; we will use the acronym RSTW. We solve RSTW by Benders' decomposition procedure (1962) as follows:

First, the binary variables, which correspond to a route sequence, are fixed at a set of values which satisfy the capacity and precedence constraints. The resulting problem is then a linear program, called the "Benders subproblem," in the scheduling variables x_i and y_i. To every linear program there corresponds another linear program called the "dual," the solutions to which are intimately related to those of the original linear program, now called the "primal." Benders' procedure calls for the solution of the dual of the Benders subproblem, which is then used to construct a constraint, called a "Benders cut," to a pure binary problem known as the "Benders master problem." The master problem is solved to produce a new set of binary variables (an improved route) which leads to a new Benders subproblem. The iterative procedure continues until the Benders master problem fails to find an improvement.

The initial route is established using a heuristic procedure to be explained later. The dual of the Benders subproblem is easily recognized as a minimum cost network flow problem. Such problems are characterized by a set of points, called "nodes," and a set of directed lines, called "arcs," connecting some pairs of nodes. Certain nodes produce flow (they are called "sources") while others demand to absorb flow (they are called "sinks"). Associated with each arc are a cost per unit flow and upper and lower bounds on the amount of flow permitted on that arc. The problem is to find the minimum total cost flow pattern which satisfies the demands of the sinks while not exceeding the supply capabilities of the sources or violating the flow bounds on any arc. Such problems occur commonly and several algorithms exist for solving general or special versions (see Ford and Fulkerson (1962), Busacker and Saaty (1965), and Klein (1967)). We devise a specialized algorithm exploiting the structure of this network so that an optimal flow pattern can be found very quickly. Using standard linear programming theory, an optimal flow pattern can be used to produce an optimal schedule for the given route sequence.

The pure binary routing problem, however, is not easily solved owing to its size and lack of clear structure. We develop a heuristic route improvement procedure which is motivated by the binary problem, but no claim can be made concerning the optimality of the final solution.

The next two sections, 3 and 4, respectively, describe the scheduling subproblem and the routing master problem. A discussion of computational experience follows in Section 5.

3. SCHEDULING SUBPROBLEM.

Suppose that the route sequence is temporarily fixed. This sequence might have been produced by the heuristic procedure (described in Section 4 below) for finding an initial route; or, it may be the result of one or more route improvement cycles (also described in Section 4); or, it may simply be an arbitrary sequence of pickups and deliveries satisfying (2) through (6). Given a route, we know the values of u_{ij}, v_{ij} and w_{ij}. Thus, in RSTW, the capacity constraint set (2) disappears, and the headway constraints (3) through (6) simplify. More importantly, we now know the values of the subscripts in parentheses in (1), and the result is a linear program in the scheduling variables x_i and y_i. This program involves choosing the times for each pickup and delivery to minimize (1) for the given route sequence.

Choi (1984) showed that this program is highly redundant and that the only headway constraints (3) through (6) which must be retained are those corresponding to adjacent tasks in the route sequence. The proof is straightforward and relies only on the triangle inequality. To simplify the program, we define the following indicators of immediate adjacency. Let h_i=1 if the pickup of load i immediately precedes the delivery of load i, and h_i=0 otherwise. Similarly, let $h_{01}(i,j)$=1 if the pickup of load i immediately precedes the delivery of load j, and equal zero otherwise. Define $h_{00}(i,j)$, $h_{10}(i,j)$ and $h_{11}(i,j)$ in similar fashion using the notation convention described in Section 2 for travel times. Fixing a route sequence, then, determines the values of all of the h-variables. The scheduling linear program may now be written as

Minimize z'

subject to

$$\begin{aligned}
y_i - x_i &\geq d'_{01}(i,i) && \text{if } h_i = 1 \\
x_j - x_i &\geq d'_{00}(i,j) && \text{if } h_{00}(i,j) = 1 \\
x_j - y_i &\geq d'_{10}(i,j) && \text{if } h_{10}(i,j) = 1 \\
y_j - x_i &\geq d'_{01}(i,j) && \text{if } h_{01}(i,j) = 1 \\
y_j - y_i &\geq d'_{11}(i,j) && \text{if } h_{11}(i,j) = 1 \\
x_i \text{ and } y_i &\text{ free} && \text{for } i=1,2,\ldots,n
\end{aligned}$$

where z' is given in (1).

We next make a few routine transformations to eliminate the absolute values in (1). For $i=1,2,\ldots,n$, let

$$x_i - x_i^0 = \alpha_i^+ - \alpha_i^-$$

and

$$y_i - y_i^0 = \beta_i^+ - \beta_i^-$$

where the variables α_i^+, α_i^-, β_i^+, and β_i^- are taken to be nonnegative. Then let

$$\alpha_i^+ + \alpha_i^- - x_i^0 + \ell_i^0 = \gamma_{i1}^+ - \gamma_{i1}^-$$

and

$$\beta_i^+ + \beta_i^- - y_i^0 + \ell_i^1 = \gamma_{i2}^+ - \gamma_{i2}^-$$

for $i=1,2,\ldots,n$. Deleting constants from the objective function and rewriting the constraints yields the linear program (P)

$$\text{Minimize } z'' = \beta_{(n)}^+ - \beta_{(n)}^- - \alpha_{(1)}^+ + \alpha_{(1)}^- + \sum_{i=1}^{n} (e_i^1 \gamma_{i1}^+ + e_i^2 \gamma_{i2}^+)$$

subject to

$$[\lambda_i] \quad \beta_i^+ - \beta_i^- - \alpha_i^+ + \alpha_i^- \geq d'_{01}(i,i) + x_i^0 - y_i^0, \; h_i = 1$$

$$[P_{00}(i,j)] \quad \alpha_j^+ - \alpha_j^- - \alpha_i^+ + \alpha_i^- \geq d'_{00}(i,j) + x_i^0 - x_j^0, \; h_{00}(i,j) = 1$$

$$[P_{10}(i,j)] \quad \alpha_j^+ - \alpha_j^- - \beta_i^+ + \beta_i^- \geq d'_{10}(i,j) + y_i^0 - x_j^0, \; h_{10}(i,j) = 1$$

$$[P_{01}(i,j)] \quad \beta_j^+ - \beta_j^- - \alpha_i^+ + \alpha_i^- \geq d'_{01}(i,j) + x_i^0 - y_j^0, \; h_{01}(i,j) = 1$$

$$[P_{11}(i,j)] \quad \beta_j^+ - \beta_j^- - \beta_i^+ + \beta_i^- \geq d'_{11}(i,j) + y_i^0 - y_j^0, \; h_{11}(i,j) = 1$$

$$[s_i] \quad - \alpha_i^+ - \alpha_i^- + \gamma_{i1}^+ - \gamma_{i1}^- = - x_i^0 + \ell_i^0 \quad i=1,2,\ldots,n$$

$$[t_i] \quad - \beta_i^+ - \beta_i^- + \gamma_{i2}^+ - \gamma_{i2}^- = - y_i^0 + \ell_i^1 \quad i=1,2,\ldots,n$$

$$\alpha_i^+, \alpha_i^-, \beta_i^+, \beta_i^-, \gamma_{i1}^+, \gamma_{i1}^-, \gamma_{i2}^+ \text{ and } \gamma_{i2}^- \geq 0 \quad i=1,2,\ldots n$$

In the preceding formulation, the dual variables are indicated in brackets. The dual problem (D) of (P), as required by Benders' procedure, is:

$$\text{(D) Maximize } W = \sum_{i=1}^{n} (d'_{01}(i,i) + x_i^0 - y_i^0) h_i \lambda_i +$$
$$\sum_{i \neq j}\sum (d'_{00}(i,j) + x_i^0 - x_j^0) h_{00}(i,j) P_{00}(i,j) +$$
$$\sum_{i \neq j}\sum (d'_{10}(i,j) + y_i^0 - x_j^0) h_{10}(i,j) P_{10}(i,j) +$$
$$\sum_{i \neq j}\sum (d'_{01}(i,j) + x_i^0 - y_j^0) h_{01}(i,j) P_{01}(i,j) +$$
$$\sum_{i \neq j}\sum (d'_{11}(i,j) + y_i^0 - y_j^0) h_{11}(i,j) P_{11}(i,j) -$$
$$\sum_i (x_i^0 - \ell_i^0) s_i - \sum_i (y_i^0 - \ell_i^1) t_i$$

subject to

$$[\alpha_i^-],[\alpha_i^+] \quad - \Delta_i^0 - s_i \leq - h_i \lambda_i - \sum_{j \neq i} h_{00}(i,j) P_{00}(i,j) +$$
$$\sum_{j \neq i} h_{00}(j,i) P_{00}(j,i) + \sum_{j \neq i} h_{10}(j,i) P_{10}(j,i) - \sum_{j \neq i} h_{01}(i,j) P_{01}(i,j)$$
$$\leq - \Delta_i^0 + s_i \quad i=1,2,\ldots,n$$

$$[\beta_i^-],[\beta_i^+] \quad \Delta_i^1 - t_i \leq h_i \lambda_i - \sum_{j \neq i} h_{10}(i,j) P_{10}(i,j) +$$
$$\sum_{j \neq i} h_{01}(j,i) P_{01}(j,i) - \sum_{j \neq i} h_{11}(i,j) P_{11}(i,j) + \sum_{j \neq i} h_{11}(j,i) P_{11}(j,i)$$
$$\leq \Delta_i^1 + t_i \quad i=1,2,\ldots n$$

$$[\gamma_{i1}^-],[\gamma_{i1}^+] \quad 0 \le s_i \le e_i^1 \quad i=1,2,\ldots,n$$

$$[\gamma_{i2}^-],[\gamma_{i2}^+] \quad 0 \le t_i \le e_i^2 \quad i=1,2,\ldots,n$$

$$P_{00}(i,j),\ P_{01}(i,j),\ P_{10}(i,j),\ P_{11}(i,j) \ge 0 \quad \text{for all } i \ne j$$

$$\lambda_i \ge 0 \quad i=1,2,\ldots,n$$

where $\Delta_i^0 = 1$ if load i is the first load picked up and equals zero otherwise, and where $\Delta_i^1 = 1$ if load i is the last load delivered and equals zero otherwise. (The bracketed variables are the original primal ones.)

The dual (D) may be interpreted as a minimum cost network flow problem in the following manner. Construct a network with 2n+3 nodes, one for each pickup and delivery, a source node S, a sink node T, and a transshipment node R (see example in Figure 1). Denote the node corresponding to the pickup of load i by "+i", and that corresponding to the delivery of load i by "-i". Next, add an arc from S to +(1), the first pickup, and from -(n), the last delivery, to T. For each i, add an edge (undirected arc) between R and +i and between R and -i. Finally add an arc from task α to task β if task α immediately precedes task β in the route sequence.

The variable λ_i is the flow on the arc from +i to -i, if this arc exists, while a P-variable is the flow on the arc from the first of its tasks to the second. All of these arcs are uncapacitated (they can handle any flow level) and have unit flow costs equal to the negatives of their objective function coefficients in (D). The flows from S to +(1) and from -(n) to T are forced equal to unity with zero unit cost. The capacity of the edge between R and +i equals s_i and has unit flow cost equal to $x_i^0 - \ell_i^0$; similarly, the capacity of the edge between R and -i equals t_i and has unit flow cost equal to $y_i^0 - \ell_i^1$. Figure 1 shows the network for n=3 and the route sequence +1, +2, -1, +3, -3, -2.

We could, of course, solve this problem with any of several available algorithms (see, for example, Ford and Fulkerson (1962)

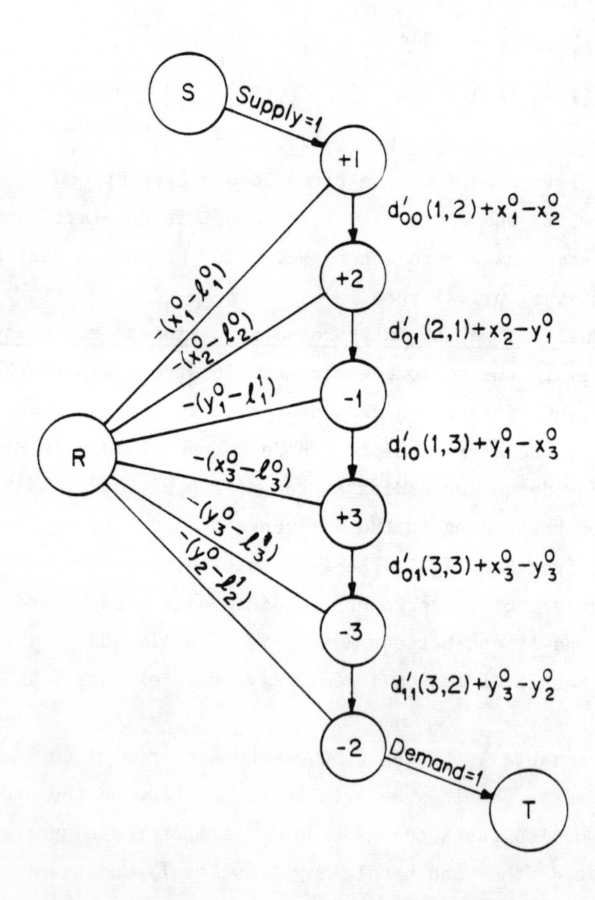

FIGURE 1. Formulation of (D) as a Network Flow Problem. Profits Per Unit Flow, e.g., $-(x_1^0 - \ell_1^0)$ and $d'_{01}(2,1) + x_2^0 - y_1^0$, are Shown Next to Each Arc and Edge.

and Klein (1967)). These are general algorithms, however, and cannot exploit special structures like the one at hand. Because our overall algorithm requires several passes through the scheduling algorithm, speed is of considerable value and therefore we developed a specialized approach for this particular network structure.

The details of the algorithm are rather intricate and we refer the reader to Choi (1984) for a complete description. The algorithm operates on the adjustment network of the given network (see Busacker and Saaty (1965)), using a number of results proven by Choi (1984) to significantly streamline the optimization process.

Once (D) is solved, it is a straightforward matter to construct an optimal solution to the scheduling subproblem using complementary slackness. In fact, Benders' procedure utilizes the solution directly so that the actual schedule needs to be produced only once at the end, when no more route improvements can be found.

4. THE ROUTING PROBLEM.

The routing problem is the "master problem" in the Benders terminology. Let m be the number of new routes already produced at some point by the route improvement procedure described below. The optimal flow pattern for each of these routes, as well as for the initial route, will have been found using the scheduling flow algorithm discussed in the previous section. The routing problem at this point thus contains m+1 constraints ("Benders cuts" in the Benders terminology), each of which has the same structural form as the objective function of the scheduling subproblem. Specifically, the routing problem is

Minimize z_1

subject to

$$
\begin{aligned}
z_1 \geq & \sum_i (d'_{01}(i,i) + x_i^0 - y_i^0)\lambda_i^{(k)} h_i \\
& + \sum_{i \neq j}\sum (d'_{00}(i,j) + x_i^0 - x_j^0)P_{00}^{(k)}(i,j)h_{00}(i,j) \\
& + \sum_{i \neq j}\sum (d'_{10}(i,j) + y_i^0 - x_j^0)P_{10}^{(k)}(i,j)h_{10}(i,j) \\
& + \sum_{i \neq j}\sum (d'_{01}(i,j) + x_i^0 - y_j^0)P_{01}^{(k)}(i,j)h_{01}(i,j) \\
& + \sum_{i \neq j}\sum (d'_{11}(i,j) + y_i^0 - y_j^0)P_{11}^{(k)}(i,j)h_{11}(i,j) \\
& - \sum_i (x_i^0 - \ell_i^0)s_i^{(k)} - \sum_i (y_i^0 - \ell_i^1)t_i^{(k)} \qquad k=1,\ldots,m+1 \\
& h \in \Omega'
\end{aligned}
$$

where the superscript (k) refers to the values of the flow variables in the corresponding scheduling network, and where Ω' is the set of all routes which satisfy the precedence and capacity constraints. A solution to this problem produces a new set of h-variables which is equivalent to a new route sequence.

Owing to the complexity of the master problem, no attempt is made to solve it exactly. Instead it is used to guide a route improvement procedure based on a Lagrangean relaxation in which the Benders cuts are dualized with equal weights into the objective function. (Computational experiments with radically different weighting patterns showed little effect on the final solution so that, given the heuristic nature of the solution procedure overall, we felt it unnecessary to search for optimal multipliers).

The route improvement procedure stems from the following idea. Each Benders cut involves only those h-variables which represent arcs present in the flow problem which produced the Benders cut. The coefficient of an h-variable in a Benders cut is the product of an arc-specific constant and the optimal flow on that arc. Our Lagrangean relaxation of the routing problem forms a new objective function, to be minimized, which is a linear combination of these cuts, while the constraint set simply guarantees route feasibility. Intuitively, then, a good solution to this relaxation will be a feasible route which tends to include arcs with small (highly negative) coefficients while avoiding those with large positive coefficients.

At any iteration, the route improvement procedure ranks from high to low the coefficients of the current objective function, which includes terms for all arcs which have previously appeared. The procedure then determines which current arc lies nearest the top of the list and attempts to eliminate it from the route. This involves moving one or both of the task nodes at the ends of this arc so as to make these two tasks nonadjacent in the route sequence. This is accomplished by first moving the task at the foot of the chosen arc into each of its feasible positions, keeping all other tasks fixed, and computing the optimal schedule for each candidate new route sequence. Then, holding this task in its locally optimal position, the task at the head of the chosen arc is moved in the same way. If both tasks end up in their original positions, then the process is repeated for the current arc with the second greatest coefficient until no further improvements can be found. If one or both tasks end up in different positions, then this sequence becomes the latest best route found and the Benders procedure returns to form a new cut, a new relaxation, and a new listing of coefficients.

There remains the issue of forming the initial route sequence with which to initiate the Benders process. We could, of course, simply pick up the first load and deliver it, pick up the second load and deliver it, and so on, using any arbitrary ordering of the loads. However, this is likely to produce a poor route sequence and the computational costs associated with starting with a poor sequence are likely to be quite high. For this reason, it is advantageous to design a heuristic procedure to obtain a good (near-optimal) initial route.

Consider the unit costs on the task-to-task arcs that appear as the right-hand-sides of the constraints in (P). Each is the sum of the travel time between the tasks (including a dwell time) and the difference between the midpoints of the task windows.

The first of these summands is called the spatial separation from the first task to the second task, while the second term is called the temporal separation from the first task to the second task. The sum is called the space-time separation from the first task to the second task. (Observe that both temporal separations and, consequently, space-time separations, may be negative.) Space-time separations very similar to these were first used by Sexton (1979) for the dial-a-ride problem. Also see Sexton and Bodin (1985a, 1985b).

Intuitively, when a space-time separation is large, we expect that the second task is unlikely to be a good successor to the first task in the route sequence. The initial routing heuristic, which we call the space-time heuristic, proceeds in the manner of the nearest neighbor algorithm for the traveling salesman problem using the space-time separations as costs. It begins by placing the pickup with the earliest pickup window midpoint at the start of the route and then, at each iteration, selects the task with the smallest space-time separation from the current last task on the partial route among all candidates for the next task. The corresponding task is added at the end of the partial route and a new iteration begins until there are no candidates for the next task. At this point, the initial route is complete.

We now state the complete algorithm for the single-vehicle routing and scheduling problem with time windows, partial loads and dwell times (see Figure 2), followed by a description of the repositioning subroutine referred to in Steps 6 and 7 (see Figure 3).

Algorithm for the RSTW:

Step 1: (Initial route) Find an initial feasible route using the space-time heuristic algorithm. This forms the incumbent route. Set K=1 and go to Step 2.

Step 2: (Optimal flow pattern for this route) Determine an optimal flow pattern for this route using the specialized network flow algorithm. Go to Step 3.

Step 3: (Lagrangean relaxation) Form the objective function of the Lagrangean relaxation of the routing problem by equally

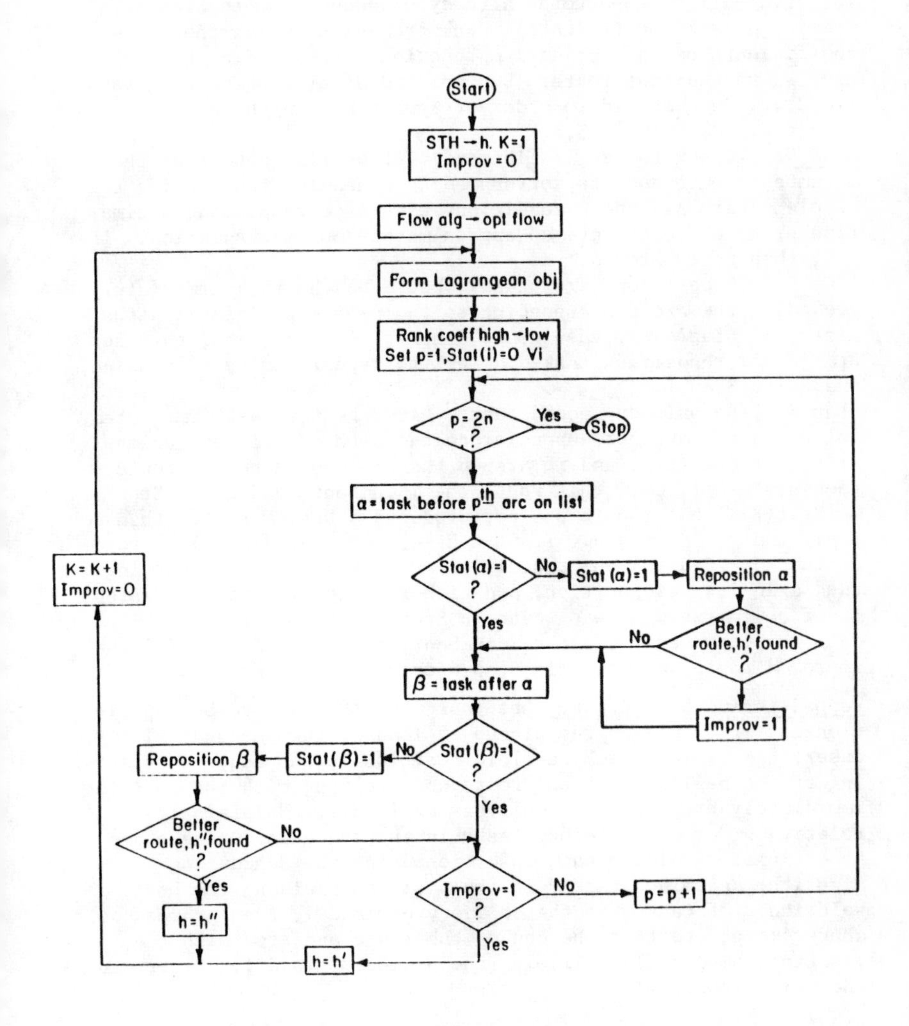

FIGURE 2. Flowchart for the RSTW Algorithm.

Key: h is the Incumbent Route Sequence.

STH is the Space-time Heuristic Algorithm.

weighting all flow patterns already produced. Go to Step 4.
Step 4: (Rank coefficients) Rank from high to low the L=2n-1 coefficients of this objective function corresponding to the L arcs in the current route. Set p=1 and Stat(i)=0 for the tasks i=1,2,...,2n numbered in order of appearance in the route sequence. Go to Step 5.
Step 5: (Check to see if finished) If p=L+1, then stop; the incumbent route and its corresponding schedule, found using the complementary slackness conditions with a corresponding optimal flow pattern, constitute an approximately optimal solution. If $p \leq L$, then go to Step 6.
Step 6: (Reposition first task) Let α be the task immediately preceding the arc corresponding to the p-th coefficient on the list. If Stat(α)=1, then go to Step 7. If Stat(α)=0, then set Stat(α)=1, reposition task α using the repositioning subroutine, and go to Step 7.
Step 7: (Reposition second task) Let β be the task immediately following the arc corresponding to the p-th coefficient on the list. If Stat(β)=1 and if Step 6 has produced a better route, then set K=K+1, call this route the incumbent, and go to Step 3. If Stat(β)=1 but Step 6 has not produced a better route, then set p=p+1 and go to Step 5. If Stat(β)=0, then set Stat(β)=1, reposition task β using the repositioning subroutine while keeping task α in its new position, and either set p=p+1 and go to Step 5 (if no improvement was found in Steps 6 and 7) or set K=K+1, call the improved route the incumbent, and go to Step 3 (if an improved route was found in Step 6 or 7).

Repositioning Subroutine: Let task γ be the task to be repositioned. If task γ is the pickup of load i, then sequentially insert task γ into each feasible position in the sequence starting at the beginning of the route and stopping with the position immediately preceding the delivery of load i. Maintain the relative order of all other tasks in the sequence during these insertions. Evaluate each insertion using the scheduling algorithm and place task γ in the position yielding the best solution. If task γ is the delivery of load i, then proceed as above except start at the end of the route and stop with the position immediately following the pickup of load i. Return to the main algorithm.

5. COMPUTATIONAL EXPERIENCE.

The algorithm described above was coded in FORTRAN for the Univac 1100 system at Stony Brook. We tested the code on a data base of six vehicles derived from the Baltimore subscriber dial-a-ride data used in Sexton (1979) and Bodin and Sexton (to appear) in the manner described below. In this section we report on the running time of this code, the effect of the route

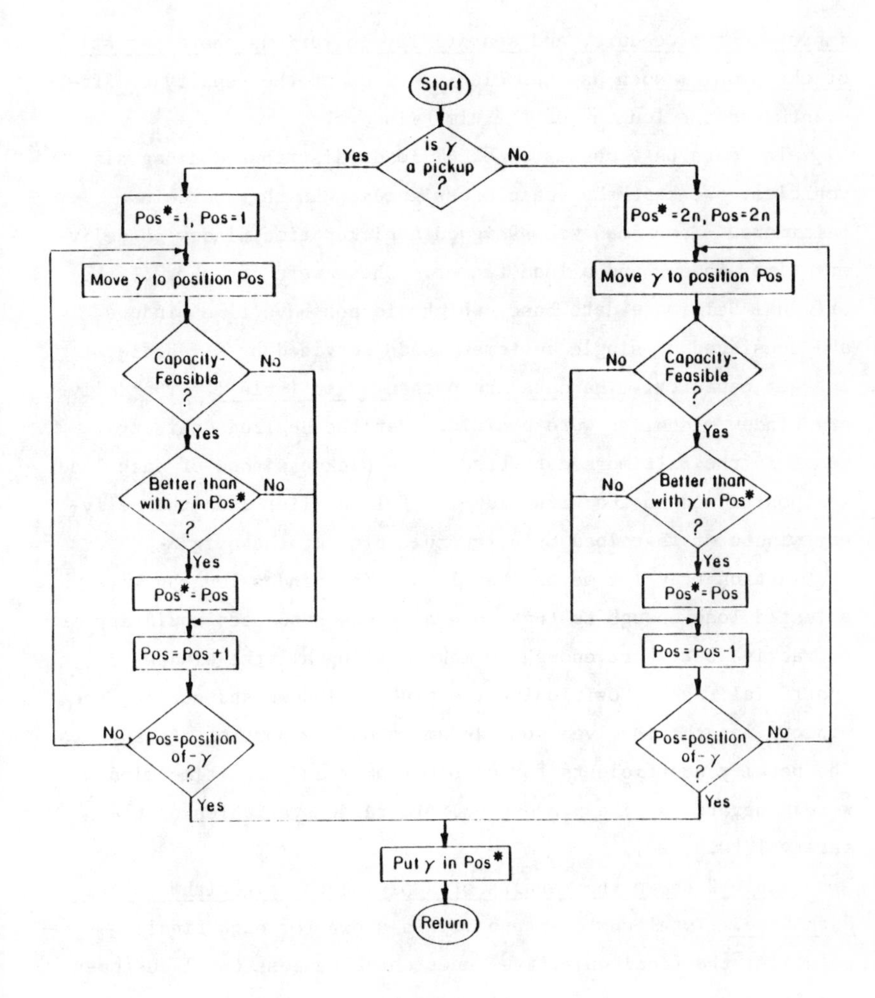

FIGURE 3. Flowchart for the Repositioning Subroutine. Key: γ is the Task to be Repositioned. If γ is the Pickup (Delivery) of Load i, Then -γ is the Delivery (Pickup) of Load i.

improvement procedure, and sensitivity to various characteristics of the problem such as the relative sizes of the penalty coefficients and the lengths of the time windows.

The data base consisted of 67 loads distributed among six vehicles. We kept the vehicle assignments as they were in Baltimore. Each load was assigned a pickup time window, a delivery time window, and a load factor. These were not part of the original Baltimore data base, which did not have time windows and consisted of single customer loads serviced by a vehicle of integer capacity. The data are presented in Table 1. The delivery window midpoints were positioned at the desired delivery times in the Baltimore data base. The pickup window of each load was positioned so that its midpoint fell earlier than the delivery window of that load by a few multiples of the origin-destination travel time of that load. The window lengths were selected long enough so that several route sequences would appear attractive but short enough to make meeting all the windows a nontrivial task. Positioning the pickup windows and setting the windows lengths involved some judgment on the part of the authors. The penalty coefficients for each customer and each time window were generated as independent uniform random variables on the interval [0.5, 20].

Table 2 shows the results of applying the algorithm to this data base. Total customer penalty is shown for each final solution; the final objective function value less total customer penalty equals the vehicle operating time. The number of iterations reported is the number of improved solutions found by the route improvement procedure. The number of scheduling algorithm calls refers to the number of times the scheduling algorithm was invoked to evaluate a route. This includes calls from within the repositioning subroutine.

We note that the initial value of the objective function was improved only marginally by the route improvement procedure. This naturally raised questions. Was it that the route improve-

TABLE 1. Data Base Used to Test the Algorithm.

Load	Origin	Dest	Time Window Pickup		Delivery		Load Factor	
18	26	25	581	586	590	600	0.05	
19	9	27	511	521	535	540	0.05	
20	30	3	348	358	380	420	0.05	Vehicle 1
21	31	3	448	454	473	475	0.05	
22	32	3	397	407	425	475	0.30	
23	31	3	443	458	474	475	0.05	
24	29	28	305	320	331	400	0.05	
77	99	98	699	708	715	720	0.05	
78	100	3	511	516	535	540	0.10	
79	101	50	488	493	499	510	0.20	
80	40	50	444	449	479	510	0.05	
81	103	50	458	464	480	510	0.20	Vehicle 2
82	102	50	439	444	480	510	0.05	
83	105	104	395	415	413	440	0.25	
84	105	104	368	375	387	440	0.05	
85	4	6	341	346	362	425	0.10	
25	34	33	697	708	715	720	0.05	
26	35	11	582	592	615	615	0.05	
27	37	36	550	555	560	570	0.05	
28	41	38	502	510	524	530	0.50	
29	40	27	476	481	490	525	0.10	Vehicle 3
30	40	39	447	452	464	510	0.20	
31	42	11	402	409	415	460	0.30	
32	44	43	381	386	396	420	0.05	
33	46	45	320	327	363	380	0.30	
34	47	45	232	242	281	300	0.10	
53	76	75	654	655	665	670	0.05	
54	77	3	522	529	540	545	0.20	
55	78	3	506	510	540	545	0.10	
56	79	34	475	481	493	500	0.05	
57	81	34	426	435	441	500	0.10	
58	80	34	449	456	468	500	0.10	Vehicle 4
59	33	34	398	406	420	500	0.05	
60	83	82	275	281	309	470	0.20	
61	84	82	258	263	310	470	0.10	
62	85	82	341	347	383	465	0.05	
63	87	86	191	210	231	390	0.10	

TABLE 1. Data Base Used to Test the Algorithm (continued).

Load	Origin	Dest	Time Window Pickup		Delivery		Load Factor	
64	5	3	693	699	715	720	0.05	
65	6	4	660	665	681	695	0.10	
66	9	7	631	636	650	670	0.20	
67	9	8	612	617	626	660	0.05	
68	13	11	540	548	572	600	0.05	
69	14	11	532	538	571	600	0.25	
70	12	11	578	588	593	600	0.05	Vehicle 5
71	88	15	466	472	490	495	0.20	
72	93	89	409	418	430	435	0.05	
73	96	90	342	346	383	420	0.05	
74	97	91	333	339	392	405	0.10	
75	95	92	271	276	301	395	0.05	
76	94	92	228	236	251	395	0.10	
1	2	1	667	680	675	690	0.05	
2	16	15	601	608	625	630	0.05	
3	9	17	583	589	601	615	0.05	
4	19	18	572	575	577	585	0.05	
5	22	3	500	515	533	565	0.05	
6	24	3	479	489	531	565	0.05	
7	24	3	491	499	531	565	0.05	
8	20	3	540	550	557	565	0.05	
9	21	3	425	431	457	470	0.50	Vehicle 6
10	71	3	376	386	403	480	0.05	
11	68	3	173	179	196	206	0.05	
12	70	3	384	399	403	480	0.05	
13	74	3	231	240	281	380	0.05	
14	73	3	245	260	280	380	0.50	
15	69	3	335	345	352	400	0.05	
16	72	3	299	315	328	350	0.45	
17	67	3	353	365	365	400	0.05	

ment procedure was ineffective, or was it that the space-time heuristic procedure was finding very good starting solutions? In his dissertation, Choi (1984) experimented with five variants of the initial routing heuristic and two variants of the route improvement procedure. He discovered that almost invariably any combination of the two produced the same final objective function value but that this initial routing heuristic and one other consistently produced the best starting solutions. A variant of the

TABLE 2. Results of the Experiments.

Veh	n	STH Obj	Final Obj	VOT VWT	CP TWDev #WM	# Iter	Sched Alg Calls	CPU (sec)
1	7	274.6	274.3	273.0 72.8	1.3 1.0 2	1	106	1.1
2	9	366.1	366.0	346.8 148.8	19.2 22.2 2	1	182	3.4
3	10	486.6	472.7	470.2 115.6	2.5 2.8 1	2	271	7.5
4	11	465.8	454.5	452.3 170.4	2.2 2.7 1	1	270	7.9
5	13	471.3	468.8	452.9 82.1	15.9 26.5 1	1	354	15.8
6	17	503.9	503.9	503.9 122.9	0.0 0.0 0	0	426	24.7

Key:
STH=Space-Time Heuristic VOT=Vehicle Operating Time
VWT=Vehicle Waiting Time CP=Customer Penalty
TWDev=Time Window Deviation #WM=Number of Windows Missed

initial routing procedure produced objective function values of 2097.6, 3684.3, 3524.7, for vehicles 4, 5, and 6, respectively; total CPU time for these starting solutions were 20.0, 34.8, and 91.3 seconds, respectively. We conclude, therefore, that this initial routing heuristic does perform well in general and that the route improvement procedure based on the Benders' decomposition is sufficiently powerful to find better solutions when given a poor initial solution, but at a considerable cost in computation time.

We observe that the CPU time is quite reasonable. This is

important if such a procedure is to be used in a multi-vehicle algorithm where one may expect large CPU time requirements for the customer assignment routine. Based on the experience of Bodin and Sexton (to appear) with the dial-a-ride problem, we are optimistic that a multi-vehicle algorithm based on this single-vehicle algorithm can be made to run acceptably fast for practical applications.

The space-time heuristic algorithm is a one-pass procedure and we can be confident that its CPU requirements are a small fraction of the total CPU time. This may tempt one into eliminating the route improvement procedure and proceeding strictly on the basis of the initial solution. Based on the experience of Sexton (1979) and Sexton and Bodin (1985a, 1985b) with a similar space-time approach, this would not be well-advised. We believe that the performance of the space-time procedure depends rather heavily on the problem structure, in ways we cannot yet predict, and that the results reported here may not be typical of its behavior in general. Therefore, we recommend the use of the space-time heuristic algorithm only in conjunction with some form of post-heuristic improvement procedure.

We would expect total customer penalty, total vehicle operating time, and total vehicle waiting time (the portion of total vehicle operating time during which the vehicle is neither traveling nor loading or unloading at a site) to depend on the values of the penalty coefficients e_i^1 and e_i^2. We used the data from vehicle 1 in the above data base, set $e_i^1 = e_i^2 = p_c$ for all i, and executed algorithm RSTW with p_c = 0.01, 0.1, 1.0, and 10. The results are displayed in Table 3 which indicates that as p_c increases, (1) total time window deviation decreases, (2) total vehicle operating time increases, (3) total vehicle waiting time increases, and (4) the number of tasks with positive time window deviation decreases. Thus the algorithm does display the requisite sensitivity to its coefficients. We note in particular the abrupt alteration in the nature of the solution as p_c changes from 0.10 to 1.00: the vehicle spends about 140 more minutes "on the street", about half of it waiting, in order to miss fewer windows and incur less customer penalty. We set p_c equal to a very large positive constant (we used one billion) to try to solve the hard-window problem but the algorithm could not find a feasible solution, that is, one which satisfied all the time window and capacity constraints.

TABLE 3. Sensitivity of the Algorithm to the Coefficients.

Common Penalty (p_c)	0.01	0.10	1.00	10.00
Objective Value	135.63	178.47	273.94	282.94
Total Cust Pen	5.19	48.25	3.94	10.00
Vehicle Op Time	130.44	130.21	270.00	272.94
Total Window Dev	519.00	482.50	3.94	1.00
Vehicle Wait Time	0.00	0.00	69.87	72.81
CPU Time (sec)	2.64	3.49	0.82	0.97
No. Windows Missed	12	12	3	2

We next examined the dependence of CPU time on the widths of the time windows. We created seven test problems of various sizes from the data base, set all window widths $u_i^0 - \ell_i^0 = u_i^1 - \ell_i^1 = T$, and executed the algorithm with T = 5, 10, and 15. The results appear in Table 4. As expected, CPU time increases with both the number of loads (see also Table 2) and the tightness of the time windows.

TABLE 4. Average CPU Time (sec) as a Function of n and T.

T \ n	5	7	9	11	13	15	17
5	0.2	1.0	4.4	24.0	19.0	37.9	61.5
10	0.2	0.9	2.6	14.6	18.1	27.7	40.0
15	0.2	0.9	2.7	14.0	15.6	18.9	30.1
Average	0.2	0.9	3.2	17.5	17.6	28.2	43.9

6. CONCLUSIONS.

We have presented a formulation of the single vehicle routing and scheduling problem with time windows, partial loads and dwell times, and illustrated how Benders' decomposition procedure can be used to solve this problem approximately. The scheduling component was shown to be the dual of a network flow problem whose special structure enabled us to devise an exact algorithm for its solution. The routing component was attacked by a route improvement procedure based on the Benders master problem, a rather complex nonlinear integer program. A specialized heuristic algorithm for providing an initial feasible route was described as well.

Our computational experience with the algorithm showed it to be quite efficient for problems of moderate size. In particular, problems with 17 or fewer loads (34 or fewer pickup and delivery tasks) were solved in under 30 seconds of CPU time each. A number of experiments illustrated the effects of different penalties, time window widths, and problem sizes.

In practice there are sure to be many complicating factors, such as loads which cannot be mixed, limits on the amount of time a load may remain on the vehicle, and nonlinear penalty functions. Hence, much work remains to be done on important variants of this problem and we view this paper as a first step.

REFERENCES.

Armstrong, G.R. and Garfinkel, R.S. (1982). Dynamic programming solution of the single and multiple vehicle pickup and delivery problem with application to Dial-A-Ride. Working Paper No. 162, U. of Tennessee.

Assad, A., Ball, M., Bodin, L. and Golden, B. (1981). Combined distribution, routing and scheduling in a large commercial firm. Proceedings of the 1981 Northeast AIDS Conference (R. Pavan and P. Anderson, eds.), Boston, 99-102.

Ball, M., Golden, B., Assad, A. and Bodin, L. (1983). Planning for truck fleet size in the presence of a common carrier option. Decision Sciences, 14, 103-120.

Benders, J.F. (1962). Partition procedure for solving mixed-variables programming problems. Numerische Mathematik, 4, 238-252.

Bodin, L. and Golden, B. (1981). Classification in vehicle routing and scheduling. Networks, 11, 97-108.

Bodin, L., Golden, B., Assad, A. and Ball, M. (1983). Routing and scheduling of vehicles and crews - The state of the art. Computers and Operations Research, 10, 63-211.

Bodin, L. and Sexton, T. (to appear). The multi-vehicle subscriber Dial-A-Ride problem. The Delivery of Urban Services, TIMS Studies in the Management Sciences, 22.

Busacker, R.G. and Saaty, T.L. (1965). Finite Graphs and Networks: An Introduction with Applications. McGraw-Hill, New York.

Choi, Y.-M. (1984). The single vehicle trailer routing and scheduling problem with partial loads, time windows and dwell times. Ph.D. Dissertation, Dept. Applied Mathematics and Statistics, SUNY Stony Brook, New York.

Cullen, F.H., Jarvis, J.J. and Ratliff, H.D. (1981). Set partitioning based heuristics for interactive routing. Networks, 11, 125-144.

Fisher, M.L. and Jaikumar, R. (1981). A generalized assignment heuristic for vehicle routing. Networks 11, 109-124.

Ford, L.R. and Fulkerson, D.R. (1962). Flows in Networks, Princeton University Press, Princeton, New Jersey.

Klein, M. (1967). A primal method for minimal cost flows with applications to the assignment and transportation problems. Management Science, 14, 205-220.

Lenstra, J. and Rinnooy Kan, A. (1981). Complexity of vehicle routing and scheduling problems. Networks, 11, 221-227.

Love, R.R., Jr. (1981). Traffic scheduling via Benders' decomposition. Mathematical Programming Study, 15, 102-124.

Magnanti, T.L. (1981). Combinatorial optimization and vehicle fleet planning: Perspectives and prospects. Networks, 11, 179-214.

Psaraftis, H.N. (1980). A dynamic programming solution to the single vehicle many-to-many immediate request Dial-A-Ride problem. Transportation Science, 14, 130-154.

Psaraftis, H.N. (1983). An exact algorithm for the single vehicle many-to-many Dial-A-Ride problem with time windows. Transportation Science, 17, 351-357.

Sexton, T.R. (1979). The single vehicle many-to-many routing and scheduling problem. Ph.D. Dissertation, Dept. Applied Mathematics and Statistics, SUNY Stony Brook, New York.

Sexton, T.R. and Bodin, L.D. (1981). The single vehicle many-to-many routing and scheduling problem with customer-dependent dwell times. Working Paper, SUNY Stony Brook.

Sexton, T.R. and Bodin, L.D. (1985a). Optimizing single vehicle many-to-many operations with desired delivery times: I. scheduling. Transportation Science, 19, 378-410.

Sexton, T.R. and Bodin, L.D. (1985b). Optimizing single vehicle many-to-many operations with desired delivery times: II. routing. Transportation Science, 19, 411-435.

Solomon, M.M. (1978). Vehicle routing and scheduling with time window constraints: Models and algorithms. Working Paper No. 83-02-01, The Wharton School, U. of Pennsylvania.

Stein, D.M. (1978). Scheduling Dial-A-Ride transportation systems. Transportation Science, 12, 232-249.

Received 8/85; Revised 7/14/86.

AMERICAN JOURNAL OF MATHEMATICAL AND MANAGEMENT SCIENCES

THE MINIMUM SPANNING TREE PROBLEM WITH TIME WINDOW CONSTRAINTS

Marius M. Solomon,
Assistant Professor, Joseph G. Reisman Research Professor
College of Business Administration
Northeastern University
314 Hayden Hall
Boston, Massachusetts 02115

SYNOPTIC ABSTRACT

This paper is concerned with the computational complexity and the design and analysis of algorithms for the minimum spanning tree problem with time window constraints; such constraints alter the computational complexity of even "easy" problems involving routing components. It is shown that the minimum spanning tree problem with time windows is NP-hard. We then develop $0(n^2)$ greedy and insertion approximate algorithms for its solution. Finally, we report our computational experience with these algorithms; this experience indicates that the insertion heuristic had much better performance than the greedy.

Key Words and Phrases: minimum spanning tree; time window constraints; computational complexity; heuristics; computational performance.

1986, VOL. 6, NOS. 3 & 4, 399-421
0196-6324/86/030399-23 $7.60

1. INTRODUCTION.

In this paper we consider the minimum spanning tree problem with time window constraints (MSTPTW). In this problem it is desired to form a minimum cost spanning tree directed from a specified root node, indexed by 0, such that the time window constraints on the nodes are satisfied. Each time window is defined by the earliest time, e_i, and the latest time, ℓ_i, allowed for visiting node i.

In case of early arrival at a node, waiting is allowed. Hence, if node j is visited immediately after node i, the time at which node j is visited, v_j, is $\max\{e_j, v_i + t_{ij}\}$, where t_{ij} is the direct travel time between i and j. We assume that the cost of direct travel from node i to node j is given by the corresponding distance, d_{ij}.

Recently, we have witnessed the emergence of allowable service time constraints, or time windows, as an important area for progress in handling realistic generalizations of models involving routing components. These time windows arise naturally in many important practical problems (Bodin, Golden, Assad and Ball (1983), Solomon (1983)).

The MSTPTW has a host of applications in contexts such as: computer communication networks, military logistics, and airplane scheduling for hub-and-spoke network designs.

In a centralized computer communication network, geographically dispersed terminals must communicate with a computer host located at a central site. Direct communication between any pair of nodes involves a cost of constructing or renting the respective communication line. When the terminals impose response time constraints, or they are open during specified time intervals only, the problem of determining the minimum cost network is a MSTPTW.

The MSTPTW also occurs in military logistics where troops located at a central site must be deployed to reach strategic points within specified time intervals.

Yet another application is in airplane scheduling for hub-and-spoke network designs. The additional complication here is that the underlying network is a spider graph, i.e. all nodes except the hub are required to have degree at most two (the degree of a node is the number of edges incident to the node). Each node has two one-sided time windows specifying the earliest time of pick-ups' departure from the node, and the latest time of deliveries arrival at the node, respectively.

The original work on algorithms for the minimum spanning tree problem is due to Kruskal (1956), Prim (1957), and Dijkstra (1959). The directed spanning tree problem has been solved by Edmonds (1967). Recently, we have witnessed a flurry of activity on extensions of the basic problem, all of which are computationally intractible, i.e., NP-complete (see definition in section 2). In the capacitated problem, which is common to many centralized computer communication network design problems, a limit is imposed on the amount of traffic (e.g. messages) that each port of the computer host at the central site can handle. This problem has been studied heuristically by Kershenbaum (1974). Optimal approaches have been proposed by Chandy and Lo (1973) and Gavish (1983a), (1983b). Gavish (1982) has also tackled the degree-constrained problem. The presence of resource constraints and flow requirements is analyzed in a Lagrangian relaxation context by Shogan (1983).

The minimum spanning tree problem has been used successfully to provide lower bounds for more general routing problems such as the traveling salesman problem (Held and Karp (1970), (1971)) and the vehicle routing problem (Christofides, Mingozzi, and Toth (1981)).

This paper is concerned with the computational complexity and the design and analysis of algorithms for the minimum spanning tree problem with time window constraints. The importance of these issues rests not only with the problem at hand, but also in shedding some light on the potential usefulness of using this problem as a relaxation in approaches to more general routing problems with time windows.

The paper is organized as follows. In Section 2, it is shown that this problem class is NP-hard. We then develop $O(n^2)$ approximate algorithms for its solution in Section 3. We report our computational experience in Section 4. Finally, conclusions and some directions for future research are provided in Section 5.

2. COMPUTATIONAL COMPLEXITY

We now show that the MSTPTW is NP-hard. We begin by introducing some notions of computational complexity theory that are going to be needed in our analysis.

To characterize the computational difficulty of a combinatorial optimization problem, one usually examines the theoretical efficiency of optimal algorithms designed for its solution. A generally accepted way of measuring the theoretical efficiency of an algorithm is by its running time (i.e. the required number of elementary computational steps such as additions and comparisons) as a function of problem instance size (i.e. the number of bits, or symbols, required to encode it). An algorithm is considered "good" and the corresponding problem "well solved" if the algorithm is polynomial, i.e., its running time is $O(p(n))$, where p is a polynomial function and n is the problem size (a function $q(n)$ is said to be $O(p(n))$ if there exists a constant $c \geq 0$ such that $q(n) \leq c.p(n)$, for all $n>0$). A problem has not been "well solved" if any known algorithm for its solution requires exponential time.

A large class of problems which have not been "well solved" is known as the class of NP-complete problems. The problems in this class are characterized by the fact that if one of them could be "well solved", then all of them will be. However, since notoriously hard problems such as the traveling salesman problem are NP-complete, it is highly unlikely that these problems will be well solved. A problem is NP-hard if the existence of a polynomial time algorithm for it implies the existence of a polynomial time algorithm for all NP-complete problems.

In order to show that the MSTPTW is NP-hard, we reformulate it as a decision problem since the theory of NP-completeness deals primarily with decision problems, which require a yes/no answer. For the decision version, a MSTPTW problem instance is given by: Graph $G=(N,A)$, weight $d_a \varepsilon Z^+$ for each $a \varepsilon A$, weight t_a for each $a \varepsilon A$, weight $e_i \varepsilon Z^+$ for each $i \varepsilon N$, weight $\ell_i \varepsilon Z^+$, $\ell_i > e_i$, all $i \varepsilon N$, positive integer B (Z^+ denotes the set of positive integers). The decision version then asks the question: "Is there a directed spanning tree T for G such that the sum of the weights of the edges in T does not exceed B and such that the time window constraints are satisfied? That is, is there a directed spanning tree T for G such that $\sum_{a \in T} d_a \leq B$, and there exists a one-to-one function v: $N \rightarrow Z^+$ with $e_i \leq v_i \leq \ell_i$, and $v_j \geq v_i + t_a$ if $a \varepsilon T$, $a=(i,j)$?"

We are now in a position to prove

Theorem 1. The minimum spanning tree problem with time windows is NP-hard.

Proof: It suffices to show that some known NP-complete problem is polynomially transformable to MSTPTW. Note that a problem P′ is said to be polynomially transformable to a problem P if one can define a function f, computable in polynomial time, which maps each instance $I \varepsilon P'$ to a corresponding instance $f(I) \varepsilon P$ such that I is feasible (i.e., the question can be answered affirmatively) if and only if f(I) is feasible. In other words, if P′ polynomially transforms to P, then P′ can be considered as a special case of P; Hence P is at least as hard as P′.

To prove that the MSTPTW is NP-hard, we consider the bounded diameter spanning tree problem (BDST) which is (see Garey and Johnson (1979)) NP-complete. A BDST problem instance is given by: Graph $G=(N,A)$, weight $w_a \varepsilon Z+$ for each $a \varepsilon A$, positive integer

$D<|A|$, positive integer B. Then, this decision problem asks the question: "Is there a spanning tree T for G such that the sum of the weights of the edges in T does not exceed B and such that T contains no simple path with more than D edges?"

To show that BDST polynomially transforms to MSTPTW, given any instance I of BDST, we define the corresponding instance f(I) of MSTPTW as follows: MSTPTW has the same graph as BDST, $d_a=w_a$ and $t_a=1$ for all $a\varepsilon A$, $e_i=0$ and $\ell_i=D$ for all $i\varepsilon N$.

It is easy to see from its definition that f can be computed by a polynomial time algorithm. Hence the first condition is met. We now show that the second property of a polynomial transformation is satisfied. For this, let T be a bounded diameter spanning tree for G. Clearly, T is also a spanning tree in f(G) and moreover

$$\sum_{a\varepsilon T} d_a = \sum_{a\varepsilon T} w_a \leq B. \tag{1}$$

Since T contains no simple path with more than D edges, we have $\sum_{a\varepsilon S} t_a \leq D$, where $S \subset T, |S|<D$, and hence T satisfies the time windows. Therefore, I feasible implies f(I) feasible. Conversely, suppose f(I) is feasible and let T be the associated time feasible spanning tree for f(G). T will be a spanning tree in G, and from (1) the sum of the weights in T will not exceed B. For every simple path in T we must have: $\sum_{a\varepsilon P} t_a \leq D$, where P is a simple path, and since $t_a=1$, at most D edges of T can be included in such a path, completing the proof. Therefore, MSTPTW is NP-hard.

3. HEURISTICS

Considering the computational complexity of the MSTPTW, heuristics seem to offer the most promise for realistic size problems. We develop two types of methodologies: one stemming from the minimum spanning tree problem solution, the other based on vehicle routing and scheduling problem with time windows approaches (see Solomon (1983)).

3.1. A Greedy Heuristic. A method that suggests itself is an extension of Prim's algorithm (Prim (1957)) for the minimum spanning tree problem. Let 0 denote the root node. Starting from the set $U=\{0\}$, one recursively adds to T the shortest time feasible arc leaving U until all nodes have been added to U, and a tree has been formed. When considering an arc (i,j) for inclusion in T, $i\varepsilon T$, time window violations need only be checked at j.

While this method is optimal for the unconstrained problem, there is no guarantee of optimality in the presence of time windows. One could let the heuristic search be guided by a criterion which is the weighted combination of distance and time costs, and even the urgency of visiting a node. This latter measure expresses the time remaining until the last possible visit time for every node that could be next added to the tree. This method requires $O(n^2)$ time, where n is the number of nodes. This is because the shortest time feasible arc leaving U can be found in O(n) time and this step is performed n-1 times.

3.2. An Insertion Heuristic. A different approach for solving the MSTPTW is to insert arcs into the tree guided by a measure of cost savings, given by

$$s_{ij} = d_{0j} - d_{ij}.$$

In other words, starting with the tree $T=\{(0,i):i\varepsilon N\}$, at every step in the process one replaces the arc (0,j*) with the arc (i*,j*) for which the maximum cost savings is derived and time feasibility is preserved.

If the predecessor of i* is not 0, then one only needs to ensure that j* is time feasible. Otherwise, inserting i* between 0 and j* could potentially alter the times at which the successors to j* are visited. In Solomon (1983) we have developed very efficient conditions for testing the time feasibility of such insertions. We have implemented this method using list processing and heapsort structures, as proposed by Golden, Magnanti, and Nguyen (1977). The time complexity of this algorithm is $O(n^2)$.

After a tree T has been formed one could adapt this method to a 1-opt interchange procedure where arcs in T would be exchanged for lower cost time feasible arcs not in T.

To illustrate the procedure, consider the following 4-node Euclidean graph example. Let $d_{01}=d_{12}=1$, $d_{02}=d_{23}=1.4$, $d_{03}=2$, $d_{13}=2.2$. Take the time windows $[e_i, \ell_i]$, i=1,2,3, as: [0,2], [1,2], [2,3].

Starting from the tree $T_0=\{(0,1),(0,2),(0,3)\}$, since the maximum savings is $s_{23}=0.6$, the insertion heuristic will produce the optimal solution $T_1=\{(0,1),(0,2),(2,3)\}$ after one iteration. Starting from the set U={0}, the greedy heuristic will first add the arc (0,1) to the tree, then (U={0,1}) the arc (1,2) and, finally, (U={0,1,2}) the arc (0,3). Note that the greedy solution, $T_G=\{(0,1),(1,2),(0,3)\}$ is not optimal.

4. COMPUTATIONAL EXPERIENCE

In order to evaluate the computational capabilities of the algorithms presented, given that no benchmark problem set is available in the literature, we have developed such a set of problems.

The design of the test problem sets highlights several parameters that can effect the behavior of MSTPTW heuristics. These are geographical data and time window characteristics as described by the percentage of time constrained nodes and the tightness of the time windows.

One set of problems uses randomly generated geographical data (denote this as problem set R). Problem 8 from the standard set of routing test problems given in Christofides, Mingozzi, and Toth (1979) was used for this purpose. The other set of problems (denoted as problem set C) contains clustered problems based on Problem 12 from Christofides, Mingozzi, and Toth (1979).

Given a certain geographical data and time horizon, we have created the MSTPTW test problems by generating time windows of various widths, within the time horizon, for different percentages of customers. Note that for short time horizon problems, the time windows are more packed together, with more nodes having to be visited around the same time. This discourages the creation of long paths in the tree, hence higher deviations from the optimal unconstrained solution can be expected. Details of the method we have designed for the random generation of time window constraints can be found in Solomon (1983). All the test problems are 100 node Euclidean problems. This problem size is not limiting for the algorithms presented, as much larger problems could be solved. Travel times between nodes are taken to equal the corresponding distances. Summary characteristics of these problems are given in Table 1 and one problem is shown in full detail in Table 2.

TABLE 1: Summary Characteristics of the Problem Sets

Problem No. (R,C)	Time Horizon	Time Windows Proportion	Mean	Std. Dev.
1	100	100%	20	0
2	100	75%	20	0
3	100	50%	20	0
4	100	25%	20	0
5	100	10%	20	0
6	100	10%	40	0
7	200	100%	40	0
8	200	75%	40	0
9	200	50%	40	0
10	200	25%	40	0
11	200	10%	40	0
12	200	10%	80	0
13	100	100%	41	15
14	200	100%	61	16

TABLE 2: Test Problem R7

Node Number	X-Coordinate	Y-Coordinate	Earliest Visit Time	Latest Visit Time
1	35.00	35.00	0.00	200.00
2	41.00	49.00	138.00	178.00
3	35.00	17.00	46.00	86.00
4	55.00	45.00	147.00	187.00
5	55.00	20.00	133.00	173.00
6	15.00	30.00	147.00	187.00
7	25.00	30.00	25.00	65.00
8	20.00	50.00	145.00	185.00
9	10.00	43.00	54.00	94.00
10	55.00	60.00	63.00	103.00
11	30.00	60.00	66.00	106.00
12	20.00	65.00	104.00	144.00
13	50.00	35.00	80.00	120.00
14	30.00	25.00	109.00	149.00
15	15.00	10.00	32.00	72.00
16	30.00	5.00	83.00	123.00
17	10.00	20.00	40.00	80.00
18	5.00	30.00	30.00	70.00
19	20.00	40.00	83.00	123.00
20	15.00	60.00	104.00	144.00
21	45.00	65.00	158.00	198.00
22	45.00	20.00	140.00	180.00
23	45.00	10.00	26.00	66.00
24	55.00	5.00	72.00	112.00
25	65.00	35.00	30.00	70.00
26	65.00	20.00	160.00	200.00
27	45.00	30.00	73.00	113.00
28	35.00	40.00	13.00	53.00
29	41.00	37.00	43.00	83.00
30	64.00	42.00	125.00	165.00
31	40.00	60.00	135.00	175.00
32	31.00	52.00	160.00	200.00
33	35.00	69.00	160.00	200.00
34	53.00	52.00	63.00	103.00
35	65.00	55.00	78.00	118.00
36	63.00	65.00	83.00	123.00
37	2.00	60.00	53.00	93.00
38	20.00	20.00	143.00	183.00
39	5.00	5.00	86.00	126.00
40	60.00	12.00	79.00	119.00
41	40.00	25.00	160.00	200.00
42	42.00	7.00	79.00	119.00
43	24.00	12.00	36.00	76.00

TABLE 2: continued:

44	23.00	3.00	97.00	137.00
45	11.00	14.00	102.00	142.00
46	6.00	38.00	87.00	127.00
47	2.00	48.00	35.00	75.00
48	8.00	56.00	45.00	85.00
49	13.00	52.00	160.00	200.00
50	6.00	68.00	92.00	132.00
51	47.00	47.00	44.00	84.00
52	49.00	58.00	87.00	127.00
53	27.00	43.00	87.00	127.00
54	37.00	31.00	70.00	110.00
55	57.00	29.00	22.00	62.00
56	63.00	23.00	30.00	70.00
57	53.00	12.00	160.00	200.00
58	32.00	12.00	36.00	76.00
59	36.00	26.00	42.00	82.00
60	21.00	24.00	62.00	102.00
61	17.00	34.00	18.00	58.00
62	12.00	24.00	73.00	113.00
63	24.00	58.00	25.00	65.00
64	27.00	69.00	160.00	200.00
65	15.00	77.00	46.00	86.00
66	62.00	77.00	160.00	200.00
67	49.00	73.00	42.00	82.00
68	67.00	5.00	160.00	200.00
69	56.00	39.00	93.00	133.00
70	37.00	47.00	81.00	121.00
71	37.00	56.00	138.00	178.00
72	57.00	68.00	84.00	124.00
73	47.00	16.00	45.00	85.00
74	44.00	17.00	63.00	103.00
75	46.00	13.00	77.00	117.00
76	49.00	11.00	93.00	133.00
77	49.00	42.00	41.00	81.00
78	53.00	43.00	59.00	99.00
79	61.00	52.00	86.00	126.00
80	57.00	48.00	108.00	148.00
81	56.00	37.00	160.00	200.00
82	55.00	54.00	78.00	118.00
83	15.00	47.00	23.00	63.00
84	14.00	37.00	21.00	61.00
85	11.00	31.00	155.00	195.00
86	16.00	22.00	143.00	183.00
87	4.00	18.00	160.00	200.00
88	28.00	18.00	18.00	58.00
89	26.00	52.00	160.00	200.00
90	26.00	35.00	9.00	49.00
91	31.00	67.00	32.00	72.00

TABLE 2: continued:

92	15.00	19.00	43.00	83.00
93	22.00	22.00	18.00	58.00
94	18.00	24.00	160.00	200.00
95	26.00	27.00	26.00	66.00
96	25.00	24.00	57.00	97.00
97	22.00	27.00	56.00	96.00
98	25.00	21.00	61.00	101.00
99	19.00	21.00	71.00	111.00
100	20.00	26.00	159.00	199.00
101	18.00	18.00	160.00	200.00
999	0.00	0.00	0.00	0.00

The heuristics described in Section 3 have been programmed in Fortran, (a listing is given in the Appendix) and tested on the problem sets using a VAX11/780. A comparison of the two algorithms with respect to solution quality and running time is presented in Table 3.

As can be seen from these tables, the insertion heuristic uniformly performed better on both the solution quality and running time dimensions. Additional support for the effectiveness of the insertion heuristic is provided by comparing its solutions to the lower bound derived by relaxing the time window constraints and solving the minimum spanning tree problem optimally. Note that this optimal solution is 562 for R and 417 for C. This comparison reveals that this heuristic provides excellent quality solutions. Additional computational experiments conducted on problems with the same geographical data, but different time horizons, lead to the same conclusions.

Our results have an intuitively appealing flavor. The greedy heuristic initially adds arcs that are close to the root node and, because of time window violations, it is forced to make expensive inclusions at later stages. On the other hand, the insertion heuristic starts out with nodes distant from the root and close to each other, hence time window violations lead to less costly insertions in the later stages of the procedure.

TABLE 3: Relative Performance of the Algorithms

Problem	INSERTION Distance	INSERTION (% over Lower Bound)	INSERTION CPU Time	GREEDY Distance	GREEDY (% over Lower Bound)	GREEDY CPU Time
		Problem Set R.				
1	704	(25.3)	3.6	783	(39.3)	6.0
2	652	(16.0)	3.6	832	(48.0)	6.0
3	625	(11.2)	3.5	737	(31.1)	6.2
4	579	(3.0)	3.7	723	(28.6)	6.0
5	572	(1.8)	3.8	627	(11.6)	6.8
6	572	(1.8)	3.8	576	(2.5)	7.1
7	670	(19.2)	3.7	737	(31.1)	6.3
8	639	(13.7)	3.6	764	(35.9)	6.4
9	621	(10.5)	3.7	735	(30.8)	6.6
10	576	(2.5)	3.8	693	(23.3)	6.5
11	572	(1.8)	3.8	597	(6.2)	7.0
12	572	(1.8)	3.9	563	(0.2)	7.0
13	590	(5.0)	3.8	721	(28.3)	6.1
14	617	(9.8)	3.6	708	(26.0)	6.2
		Problem Set C.				
1	492	(18.2)	3.3	646	(54.9)	5.8
2	462	(10.8)	3.5	680	(63.1)	5.9
3	456	(9.3)	3.3	675	(61.9)	6.3
4	437	(4.8)	3.7	574	(37.6)	6.6
5	436	(4.5)	3.4	539	(29.2)	6.9
6	435	(4.3)	3.6	502	(20.4)	7.0
7	482	(15.6)	3.3	645	(54.7)	5.8
8	464	(11.3)	3.5	682	(63.5)	6.0
9	453	(8.6)	3.3	710	(70.3)	6.2
10	435	(4.3)	3.5	525	(25.9)	6.8
11	435	(4.3)	3.4	505	(21.1)	6.8
12	435	(4.3)	3.4	502	(20.4)	6.6
13	494	(18.5)	3.5	710	(70.3)	5.8
14	461	(10.5)	3.5	657	(57.5)	6.1

5. CONCLUSIONS

This paper has focused on the computational complexity and the design and analysis of algorithms for the MSTPTW. It was shown that this problem class is NP-hard. Subsequently, a very effective and efficient insertion heuristic was developed. We believe that the better performance of the insertion heuristic relative to that of the greedy heuristic, on the test problems, is indicative of their relative performance in general.

Our computational experience also seems to indicate that the structure of the MSTPTW is more closely related to that of the vehicle routing problem with time windows than that of the minimum spanning tree. This conclusion is, at present, tentative since it is based on limited computational results conducted on routing test problems.

This problem class has opened an interesting line of research. The ideas put forth here constitute a first step toward the understanding and solution of this problem structure. Based on this framework, one could develop additional heuristics and analyze them, both computationally and theoretically, through worst-case or probabilistic methods.

Another direction for future research is the development and testing of optimal algorithms. New algorithmic possibilities created by the time window constraints are explored in Solomon (1984b), where the author describes a branch and bound algorithm. The algorithm involves the computation of lower bounds from relaxing a set of constraints which ensure that arrival at a node precedes the departure from the node. The insertion heuristic could be used to provide feasible solutions.

Given that the MSTPTW is NP-hard, it is computationally much more difficult to find an optimal time feasible spanning tree than its unconstrained counterpart. Hence, as far as using the MSTPTW in a Lagrangian relaxation context (i.e. where one dualizes a set of difficult constraints in a problem and solves the easier, relaxed problem to obtain sharp bounds on the original problem) to solve more general routing problems with time windows, it is clear that this problem does not have the same appeal that the minimum spanning tree problem has for traveling salesman and related routing problems. Nevertheless, the viability of similar approaches is a question for future research.

APPENDIX

```
C
C           ***********************
C           *  GREEDY HEURISTIC   *
C           ***********************
C
      INTEGER PRED
      COMMON/VERTEX/ETIME(200),TIMEL(200)
      COMMON/DIST/D(200,200),NNODE
      COMMON/PRED/PRED(200)
      DIMENSION INODE(200),X(200),Y(200)
C
C     DIMENSIONED FOR UP TO 200 NODES.
C     PRED(K)    = PREDECESSOR OF NODE K
C     D(I,J) = DISTANCE BETWEEN NODES I AND J
C
      INCLUDE 'DATA.INC '
      NC1=NNODE-1
      DO 30 I=1,NC1
      NC2=I+1
C     COMPUTATION OF INTER-NODE DISTANCES
      DO 40 J=NC2,NNODE
      D(I,J)=SQRT((X(I)-X(J))**2.+(Y(I)-Y(J))**2.)
      D(J,I)=D(I,J)
40    CONTINUE
30    CONTINUE
C     CALL TO THE GREEDY HEURISTIC
      CALL GREEDY
C     THE SOLUTION TREE IS OBTAINED FROM
C     THE NODE AND THE PREDECESSOR ARRAYS
      WRITE(6,2000)(I,PRED(I),I=1,NNODE)
2000  FORMAT(1X,'NODE',1X,I3,1X,'PREDECESSOR',1X,I3)
      STOP
      END
C
      SUBROUTINE GREEDY
      REAL MIN,MIN1
      COMMON/VERTEX/ETIME(200),TIMEL(200)
      COMMON/DIST/D(200,200),NNODE
      COMMON/PRED/PRED(200)
      DIMENSION TIME(200),D1(200),WTI(200),IT(200)
      INTEGER P(200),ITREE(200),PRED
C
C     TIME(K)   = VISIT TIME AT NODE K
C     WTI(K)    = WAITING TIME AT NODE K
C     NUMNOD    = NUMBER OF NODES IN THE CURRENT
C                      PARTIAL TREE
C     ITREE(K) = KTH NODE ADDED TO THE PARTIAL TREE
```

```
C
      DO 10 I=1,NNODE
      TIME(I)=0.
      D1(I)=D(1,I)
      PRED(I)=0
      IT(I)=0
      ITREE(I)=0
10    CONTINUE
      ITREE(1)=1
      CALL SORT2(NNODE,D1,P)
      ITREE(2)=P(2)
      PRED(ITREE(2))=1
      TIME(ITREE(2))=D(1,ITREE(2))
      WTI(ITREE(2))=0.
      IF(TIME(ITREE(2)).GE.ETIME(ITREE(2)))GOTO 20
      WTI(ITREE(2))=ETIME(ITREE(2))-TIME(ITREE(2))
      TIME(ITREE(2))=ETIME(ITREE(2))
20    NUMNOD=2
      COST=D(1,ITREE(2))
30    MIN1=10000000.
      DO 40 J=1,NNODE
      MIN=10000000.
      IT(J)=0
      INDEX=0
      DO 50 L=1,NUMNOD
50    IF(ITREE(L).EQ.J)INDEX=1
      IF(INDEX.EQ.1)GOTO 40
      DO 60 K=1,NUMNOD
      DKJ=D(ITREE(K),J)
      TJ=TIME(ITREE(K))+DKJ
      IF(TJ.GT.TIMEL(J))GOTO 60
      IF(TJ.LT.ETIME(J))TJ=ETIME(J)
      CRIT1=DKJ
      IF(CRIT1.GT.MIN)GOTO 60
      MIN=CRIT1
      IT(J)=K
60    CONTINUE
      CRIT2=MIN
      IF(CRIT2.GT.MIN1)GOTO 40
      ITJ=IT(J)
      JJ=J
      MIN1=CRIT2
40    CONTINUE
      NUMNOD=NUMNOD+1
      ITREE(NUMNOD)=JJ
      PRED(JJ)=ITREE(ITJ)
      COST=COST+D(PRED(JJ),JJ)
      TIME(JJ)=TIME(PRED(JJ))+D(PRED(JJ),JJ)
      IF(TIME(JJ).GE.ETIME(JJ))GOTO 80
      WTI(JJ)=ETIME(JJ)-TIME(JJ)
```

```
      TIME(JJ)=ETIME(JJ)
80    IF(NUMNOD.LT.NNODE)GOTO 30
      WRITE(6,1000)COST
1000  FORMAT(//,1X,'COST',1X,F7.2,//)
      RETURN
      END
C
      SUBROUTINE SORT2(N,A,P)
C     BUBBLE SORT
      PARAMETER(MAXPTS=200)
      DIMENSION A(MAXPTS)
      INTEGER P(MAXPTS)
      DO 10 I=1,N
10    P(I)=I
      N1=N
20    IT=0
      DO 30 I=2,N1-1
      IF(A(P(I)).LE.A(P(I+1)))GOTO 30
      IT=I
      ITEMP=P(I)
      P(I)=P(I+1)
      P(I+1)=ITEMP
30    CONTINUE
      N1=IT
      IF(N1.GT.1)GOTO 20
      RETURN
      END
C     DATA.INC
C
C     THIS ROUTINE READS IN THE PROBLEM DATA.
C
C     INODE(K) = NODE NUMBER
C     NNODE    = NUMBER OF NODES
C     X(K)     = KTH NODE X-COORDINATE
C     Y(K)     = KTH NODE Y-COORDINATE
C     ETIME(K) = KTH NODE EARLIEST VISIT TIME
C     TIMEL(K) = KTH NODE LATEST VISIT TIME
C
      OPEN(UNIT=20,FILE='R7.DAT',STATUS='OLD')
      DO 31 K=1,200
      READ(20,*)INODE(K),X(K),Y(K),ETIME(K),TIMEL(K)
      IF(INODE(K).GE.999)GOTO 32
31    CONTINUE
32    NNODE=K-1
      CLOSE(UNIT=20)
```

```
C
C          **************************
C          *  INSERTION HEURISTIC  *
C          **************************
C
      REAL D(200),SAV(40000),X(200),Y(200),
     @TIME(200),TIMEL(200),ETIME(200),WTI(200),PF(200)
      INTEGER ID(40000,2),INODE(200),IP(40000),
     @JP(40000),KPRED(200,200),NDIST(200)
      COMMON/SORT/SAV,IP,JP
C
C    D(K)         = DISTANCE FROM THE ROOT NODE TO NODE K
C    ID(II,1)     = FROM NODE OF ARC II
C    ID(II,2)     = TO NODE OF ARC II
C    SAV(II)      = SAVINGS FOR ARC II=(I,J) WHEN
C                   REPLACING ARC (1,J) WITH ARC (I,J)
C    NDIST(K)     = NUMBER OF PREDECESSORS OF NODE K
C    KPRED(I,K)   = KTH PREDECESSOR OF NODE I
C    PF(K)        = PUSH FORWARD IN VISIT TIME AT K
C
      INCLUDE 'DATA.INC'
      NARC=0
      NDIST(1)=1
      KPRED(1,1)=1
      KPRED(1,2)=0
      NC=NNODE-1
      I=1
      DO 30 J=2,NNODE
      PF(J)=0.
      NDIST(J)=1
      KPRED(J,1)=J
      KPRED(J,2)=1
      DO 40 I1=3,200
      KPRED(J,I1)=0
40    CONTINUE
      NARC=NARC+1
      ID(NARC,1)=I
      ID(NARC,2)=J
      DIST=(X(I)-X(J))**2.+(Y(I)-Y(J))**2.
      D(J)=SQRT(DIST)
      TIME(J)=D(J)
      IF(TIME(J).GE.ETIME(J))GOTO 30
      WTI(J)=ETIME(J)-TIME(J)
      TIME(J)=ETIME(J)
30    CONTINUE
      II=0
      DO 50 I=2,NNODE
      DO 60 J=2,NNODE
      IF(I.EQ.J)GOTO 60
      NARC=NARC+1
```

```
      II=II+1
      ID(NARC,1)=I
      ID(NARC,2)=J
      DIST=(X(I)-X(J))**2.+(Y(I)-Y(J))**2.
      DIST=SQRT(DIST)
C     SAVINGS COMPUTATION
      SAV(II)=D(J)-DIST
      IF(SAV(II).LT.0.)SAV(II)=0.
60    CONTINUE
50    CONTINUE
      COST=0.
      DO 70 J=2,NNODE
70    COST=COST+D(J)
      DO 80 I=1,II
      JP(I)=I
80    IP(I)=I
      CALL HPSORT(II)
90    CONTINUE
      K=1
      IF(SAV(K).LE.0)GOTO 100
      P=IP(K)+NC
      IN=ID(P,1)
      JN=ID(P,2)
      IF(NDIST(JN).NE.1)GOTO 110
      IF(NDIST(IN).EQ.1)GOTO 120
      DO 120 J=2,NDIST(IN)
      IF(KPRED(IN,J).EQ.JN)GOTO 110
120   CONTINUE
C     TIME WINDOW FEASIBILITY
      DIST=(X(IN)-X(JN))**2.+(Y(IN)-Y(JN))**2.
      DIST=SQRT(DIST)
      TJ=TIME(IN)+DIST
      IF(TJ.GT.TIMEL(JN))GOTO 110
      IF(TJ.LE.ETIME(JN))GOTO 130
      PF(JN)=TJ-TIME(JN)
      DO 140 I=2,NNODE
      IF(KPRED(I,NDIST(I)).NE.JN)GOTO 140
      IF(I.EQ.JN)GOTO 140
      DO 150 J=1,NDIST(I)-1
      IJ1=KPRED(I,NDIST(I)+1-J)
      IJ=KPRED(I,NDIST(I)-J)
      PF(IJ)=PF(IJ1)-WTI(IJ)
      IF(PF(IJ1).LE.0)PF(IJ)=0.
      IF(PF(IJ).LE.0)GOTO 150
      TNEW=TIME(IJ)+PF(IJ)
      IF(TNEW.GT.TIMEL(IJ))GOTO 110
150   CONTINUE
140   CONTINUE
      DO 160 I=2,NNODE
      IF(KPRED(I,NDIST(I)).NE.JN)GOTO 160
```

```
      IF(PF(I))170,160,180
170   WTI(I)=-PF(I)
      GOTO 160
180   TIME(I)=TIME(I)+PF(I)
      WTI(I)=0.
160   CONTINUE
      GOTO 190
130   WTI(JN)=ETIME(JN)-TJ
190   CONTINUE
      COST=COST-(D(JN)-DIST)
      DO 200 I=2,NNODE
      IF(KPRED(I,NDIST(I)).NE.JN)GOTO 200
      DO 210 J=NDIST(I)+1,NDIST(I)+NDIST(IN)
      KPRED(I,J)=KPRED(IN,J-NDIST(I))
210   CONTINUE
      NDIST(I)=NDIST(I)+NDIST(IN)
200   CONTINUE
110   SAV(K)=0.
      CALL REDOHP(II,K)
      DO 220 I=2,NNODE
      PF(I)=0.
220   CONTINUE
      GOTO 90
100   CONTINUE
      WRITE(6,1000)COST
1000  FORMAT(//,1X,'COST',1X,F10.2,//)
C     THE SOLUTION TREE IS OBTAINED FROM THE
C     NODE AND IMMEDIATE PREDECESSOR ARRAYS
      WRITE(6,1003)(I,KPRED(I,2),I=1,NNODE)
1003  FORMAT(1X,2I4)
      STOP
      END
C
      SUBROUTINE HPSORT(NN)
C     THIS ROUTINE SORTS THE ELEMENTS IN THE
C     SAVINGS ARRAY USING A HEAP SORT
      REAL SAV(40000)
      INTEGER IP(40000),JP(40000)
      COMMON/SORT/SAV,IP,JP
      IF(NN-2)70,80,10
10    CONTINUE
      NH=NN/2
      DO 20 I=1,NH
      II=NH+1-I
      COPY=SAV(II)
      IPC=IP(II)
      JPC=IPC
30    J=II+II
      IF(J-NN)40,50,60
40    IF(SAV(J+1).GT.SAV(J))J=J+1
```

```
50    IF(SAV(J).LE.COPY)GOTO 60
      SAV(II)=SAV(J)
      IP(II)=IP(J)
      JP(IP(J))=II
      II=J
      GOTO 30
60    SAV(II)=COPY
      IP(II)=IPC
      JP(JPC)=II
20    CONTINUE
70    RETURN
80    IF(SAV(1).GE.SAV(2))GOTO 70
      DT=SAV(1)
      SAV(1)=SAV(NN)
      SAV(2)=DT
      IP(1)=2
      JP(1)=2
      IP(2)=1
      JP(2)=1
      GOTO 70
      END
C
      SUBROUTINE REDOHP(NN,IC)
C     THIS ROUTINE REARRANGES THE ELEMENTS
C     OF THE HEAP WHEN ONE HAS CHANGED
      REAL SAV(40000)
      INTEGER IP(40000),JP(40000)
      COMMON/SORT/SAV,IP,JP
      II=IC
      COPY=SAV(II)
      IPC=IP(II)
      JPC=IPC
10    J=II+II
      IF(J-NN)20,30,40
20    IF(SAV(J+1).GT.SAV(J))J=J+1
30    IF(SAV(J).LE.COPY)GOTO 40
      SAV(II)=SAV(J)
      IP(II)=IP(J)
      JP(IP(J))=II
      II=J
      GOTO 10
40    SAV(II)=COPY
      IP(II)=IPC
      JP(JPC)=II
50    RETURN
      END
```

ACKNOWLEDGMENTS

The author wishes to acknowledge the Joseph G. Reisman Fund for providing partial funding for this research. The author also wishes to thank two anonymous referees for their useful comments.

REFERENCES

Bodin, L., Golden, B., Assad, A., and Ball, M. (1983). Routing and scheduling of vehicles and crews: the state of the art. Computers and Operations Research, 10, 62-212.

Chandy, K. and Lo, T. (1973). The capacitated minimum spanning tree. Networks, 3, 173-181.

Christofides, N., Mingozzi, A., and Toth, P. (1981). Exact algorithms for the vehicle routing problem, based on spanning tree and shortest path relaxations. Mathematical Programming, 20, 255-282.

Christofides, N., Mingozzi, A., and Toth, P. (1979). The Vehicle Routing Problem. In Combinatorial Optimization, N. Christofides, A. Mingozzi, P. Toth, and C. Sardi (eds.). John Wiley & Sons, Inc., New York, 315-338.

Dijkstra, E. (1959). A note on two problems in connection with graphs. Numerische Mathematik, 1, 269-271.

Edmonds, J. (1967). Optimum branchings. Journal of Research, National Bureau of Standards, 71B, 233-340.

Garey, M. and Johnson, D. (1979). Computers and Intractability: A Guide to the Theory of NP-Completeness. Freeman, San Francisco.

Gavish, B. (1982). Topological design of centralized computer networks - Formulations and Algorithms. Networks, 12, 355-377.

Gavish, B. (1983a). Formulations and algorithms for the capacitated minimal directed tree problem. Journal of the Association for Computing Machinery, 30, 118-132.

Gavish, B. (1983b). Augmented lagrangean based algorithms for solving capacitated minimal spanning tree problems. Working Paper QM-8319, Graduate School of Management, University of Rochester, Rochester, New York.

Golden, B., Magnanti, T., and Nguyen, H. (1977). Implementing vehicle routing algorithms. Networks, 7, 113-148.

Held, M. and Karp, R., (1970). The traveling salesman problem and minimum spanning trees. Operations Research, 18, 1138-1162.

Held, M. and Karp, R., (1971). The traveling salesman problem and minimum spanning trees. Part II. Mathematical Programming, 1, 6-25.

Kershenbaum, A. (1974). Computing capacitated minimal spanning trees efficiently. Networks 4, 299-310.

Kruskal, J. (1956). On the shortest spanning subtree of a graph and the traveling salesman problem. Proceedings of the American Mathematical Society, 7, 48-50.

Prim, R. (1957). Shortest connection networks and some generalizations. Bell Systems Technical Journal, 36, 1389-1401.

Shogan, A. (1983). Constructing a minimal-cost spanning tree subject to resource constraints and slow requirements. Networks, 13, 169-190.

Solomon, M. (1983). Algorithms for the vehicle routing and scheduling problem with time window constraints. Operations Research (forthcoming).

Solomon, M. (1984a). On the worst-case performance of some heuristics for the vehicle routing and scheduling problem with time window constraints. Networks (forthcoming).

Solomon, M. (1984b). Vehicle Routing and Scheduling with Time Window Constraints Models and Algorithms. Ph.D. Dissertation, Department of Decision Sciences, The Wharton School, University of Pennsylvania, Philadelphia, Pennsylvania.

Received 4/85; Revised 3/19/86.

AMERICAN JOURNAL OF MATHEMATICAL AND MANAGEMENT SCIENCES

Book Review of
THE HIDDEN GAME OF BASEBALL: A REVOLUTIONARY APPROACH TO BASEBALL AND ITS STATISTICS
by John Thorn and Pete Palmer
Doubleday & Company
Garden City, New York, 1984, x + 419 pages, $17.95

Bruce L. Golden
College of Business and Management
University of Maryland
College Park, Maryland 20742

Edward A. Wasil
Kogod College of Business Administration
American University
Washington, D.C. 20016

Many of us who take our baseball seriously enjoy playing the "hidden game" every bit as much as (if not more than) the actual game. The hidden game is played with statistics. It does not require a baseball field but can be played anywhere (even with a beer in hand). A wide variety of baseball records have been kept for the last hundred years and they now fill many volumes. These numbers have been used over the years to evaluate and compare performances. Who was the better strikeout pitcher--Nolan Ryan or

Key Words and Phrases: baseball statistics; simulation; data analysis.

Walter Johnson? Who was the true home-run king--Hank Aaron or Babe Ruth? What would their career marks look like if Ted Williams had played for the New York Yankees and Joe DiMaggio for the Boston Red Sox? Until recently, the hidden game was played using large doses of subjective interpretation of available offensive and defensive statistics.

In The Hidden Game of Baseball, Thorn and Palmer examine the game of baseball and a host of new statistical measures in order to develop fresh insights into our national pastime. This more sophisticated and scientific approach to the analysis of baseball aims at addressing in a logical, rather than emotional, manner the above-mentioned and many other related questions.

The book, which reads more like a statistics book than one on baseball, consists of 15 chapters and requires careful reading. There are 150 pages of summary statistical tables and a four page bibliography at the end of the book. (A revised soft cover edition of the book published in 1985 updates the summary tables through 1984 and includes complete player data for that year.) Overall, the book is well-written, complete, extremely informative, and intellectually stimulating. In this book, the new breed of baseball statisticians called sabermetricians (derived from the acronym of the Society for American Baseball Research and combining it with a suffix indicating measurement), for whom this book has primarily been written, have their bible.

After an introductory Chapter 1, one fundamental question is addressed in each of the remaining 14 chapters. These questions are:

Ch. 2: What's wrong with traditional statistical measures, such as Runs Batted In, Batting Average, Slugging Percentage, Earned Run Average, Fielding Percentage, and others?

Ch. 3: What are some of the new statistical measures and where did they come from?

Ch. 4: What is the Linear Weights System for measuring proficiency in batting, fielding, base stealing, and pitching?

Ch. 5: How can one measure park impact?

Ch. 6: How can one make intelligent cross-era comparisons?

Ch. 7: How do players of today compare with those of yesteryear?

Ch. 8: Does conventional baseball wisdom (with regard to the sacrifice bunt, the steal, an intentional base on balls, the batting order, etc.) make sense?

Ch. 9: Is it possible to measure "clutch" performance?

Ch. 10: How can one measure pitching effectiveness?

Ch. 11: How can one measure fielding effectiveness?

Ch. 12: "What does it take" for a team to win the pennant?

Ch. 13: What single-season performances were the greatest?

Ch. 14: Who are the best players of all time?

Ch. 15: Who should and shouldn't be "in" the Baseball Hall of Fame?

Throughout the book, an underlying theme is that what counts is runs scored on offense and runs allowed on defense. For example, in an article in Operations Research, George Lindsey (1963) proposed the additive formula

$$\text{Run Average} = \frac{(.41)\ 1B + (.82)\ 2B + (1.06)\ 3B + (1.42)\ HR.}{AB} \quad (1)$$

Lindsey's formula (1) expresses all hits in terms of runs. The coefficients of (1) are based on 373 games played during the 1959-60 season. Note that the coefficients indicate that a home run (coefficient 1.42) is not worth as much, in terms of run production, as four singles ((.4) x (.41) = 1.64)--as is assumed in the slugging percentage. The slugging percentage, the total number of bases divided by the number of at bats, indicates that a home run is worth exactly four singles. Lindsey's method forms the basis for the Linear Weights System devised by the present book's authors.

Another example of the new statistical measures is due to Bill James (1984). In its most basic form, his measure reads:

$$\text{Runs Created} = \frac{(\text{Hits} + \text{Walks}) \times \text{Total Bases.}}{\text{At Bats} + \text{Walks}} \qquad (2)$$

"Runs Created" is a rough predictor of offensive performance for an individual, but is usually quite accurate for a team or a league. In 1983, actual runs in the American and National Leagues were 10,177 and 7,993. Using James' formula (2), runs created in the American and National Leagues become 10,192 and 7,944.

Thorn and Palmer are at their best when they discuss the history and shortcomings of each statistical measure. The arguments presented are astute, comprehensive, and logically compelling.

In addition to the compilation of statistics developed by others, Thorn and Palmer present and advocate the use of their Linear Weights System. In brief, the authors have simulated all major-league games played since 1901 and have developed a formula based on the simulation results that, in the case of batting, takes every offensive event and examines its average impact. For example, runs are related to singles (1B), doubles (2B), triples (3B), and home runs (HR) by the formula

$$\text{Runs} = (.46)\ 1B + (.80)\ 2B + (1.02)\ 3B + (1.40)\ HR + \text{residual terms.} \qquad (3)$$

Residual terms include offensive events such as base on balls, hit by pitch, stolen bases, caught stealing, and others. This particular formula measures the impact of a player's accomplishment beyond the average. In the case of the 1983 American League batting champion, Wade Boggs, the authors find that Runs = 56.9, which means that Boggs contributed 56.9 more runs than what an average player might have. When compared with eighteen other offensive statistics, the authors indicate that the Linear Weights System is the most accurate; James' runs created formula (2) ranks seventh. The Linear Weights System can also be used to measure a player's fielding, pitching, and base stealing performances.

The Linear Weights System is described in Chapter 4 and is used in every subsequent chapter to address that chapter's fundamental question. In our opinion, the book's major weakness is that the description of the simulation experiment underlying the Linear Weights System is brief and cryptic. The authors should have provided many more details as to how the simulation was actually performed.

In addition, there are a few minor complaints that one might raise. The authors mention work by Ward Larkin that uses the standard deviation rather than a simple ratio of individual (say batting average) to league norm. Why was this approach not pursued? For some reason, no reference is included in the bibliography. Also, we point out that the calculation of the Normalized ERA is not correct. The authors report that Early Wynn had a 3.20 ERA in 1950 and the league had a 4.58 ERA. They claim that Wynn performed 43% better than the league average, when in fact he performed 30% better than the league average.

In summary, we find that the authors have provided for the informed follower of baseball a book that is insightful, thought-provoking, and enjoyable. The book is also well-suited for someone with a general interest in both sports and the statistical analysis of data. The authors should be commended for not only spending a great deal of time in accumulating massive amounts of data (with the invaluable help of a computer), but for also interpreting the resulting statistical measures.

Other books offering statistical insights into baseball have appeared in the past. For example, Percentage Baseball by Earnshaw Cook (1964) caused quite a stir by suggesting a number of controversial strategies. One suggestion was to realign the batting sequence in descending order of ability. No other "baseball" book, however, has been as ambitious or comprehensive in scope as this book by Thorn and Palmer.

If the fact that Gabby Hartnett was judged by Thorn and Palmer to be the catcher of the century whets your appetite, then this is a book worth reading.

REFERENCES

Cook, E. (1964). Percentage Baseball, Waverly Press, Baltimore.

James, B. (1984). The Bill James Baseball Abstract, Ballantine Books, New York.

Lindsey, G. (1963). An investigation of strategies in baseball. Operations Research, 11, 477-501.

Received 9/16/84; Revised 12/30/85.